Sustainable Agriculture and Farming

Sustainable Agriculture and Farming

Edited by
Logen Robinson

Larsen & Keller
www.larsen-keller.com

Sustainable Agriculture and Farming
Edited by Logen Robinson
ISBN: 978-1-63549-000-8 (Hardback)

© 2017 Larsen & Keller

▤ Larsen & Keller

Published by Larsen and Keller Education,
5 Penn Plaza,
19th Floor,
New York, NY 10001, USA

Cataloging-in-Publication Data

Sustainable agriculture and farming / edited by Logen Robinson.
    p. cm.
Includes bibliographical references and index.
ISBN 978-1-63549-000-8
1. Sustainable agriculture--Textbooks. 2. Agriculture-- Textbooks. 3. Farms--Textbooks.
I. Robinson, Logen.
S494.5.S86 S87 2017
630--dc23

The publisher's policy is to use permanent paper from mills that operate a sustainable forestry policy. Furthermore, the publisher ensures that the text paper and cover boards used have met acceptable environmental accreditation standards.

Printed and bound in the United States of America.

For more information regarding Larsen and Keller Education and its products, please visit the publisher's website www.larsen-keller.com

Table of Contents

Permissions

Index

Preface

Sustainable agriculture is expanding at a rapid pace due to the growing environmental concerns. The very aim of this field is to make agriculture economically viable, socially supportive and ecologically sound. It elucidates the varied changes in the practices of farming and addresses the environmental issues arising from the unhealthy practices. This book is appropriately detailed in order to benefit the readers. It will broaden the scope of knowledge in the field of sustainable agriculture and elaborate the most significant topics related to it.

Given below is the chapter wise description of the book:

Chapter 1- Sustainable agriculture is an emerging field of study; the following chapter will not only provide an overview, it will also delve deep into the variegated topics related to it. This chapter will elaborate the various aspects of this field in a lucid manner.

Chapter 2- This chapter carefully elaborates the basic concepts of sustainable agriculture to provide a complete understanding of sustainable agriculture. Concepts like crop rotation, monoculture, polyculture, cover crop, weed control among others are described in the following chapter in a critical and systematic manner.

Chapter 3- Due to environmental concerns and depletion of natural resources, it is extremely important to practice sustainable agriculture globally. The following chapter will give a glimpse of the popular practices that are prevalent for example, windbreak, compost, no-till farming, afforestation, etc. which will help readers broaden their spectrum of knowledge.

Chapter 4- This chapter is an introduction to farming incorporating all the major aspects related to the area of study. The chapter unfolds its crucial aspects in a critical yet systematic manner. It will not only give a detailed account of farming but it will also, discuss at length the related dimensions of it like the various types of farms, equipments used for farming across the globe, etc.

Chapter 5- Farming can be further classified into individual branches of study like organic farming, intensive, extensive farming, etc. which are listed in the following chapter. These distinct types of farming are dealt with great detail in this chapter so that it provides the readers with a comprehensive account of this vast field.

Chapter 6- The aim of this chapter is to provide the readers an in-depth understanding of sustainable agriculture by elucidating the fields that are closely related to it and play a role in the field's progress. Topics such as agroecology, agroforestry, conservation agriculture among others have been rigorously examined in this chapter.

Indeed, my job was extremely crucial and challenging as I had to ensure that every chapter is informative and structured in a student-freindly manner. I am thankful for the support provided by my family and colleagues during the completion of this book.

Editor

Introduction to Sustainable Agriculture

Sustainable agriculture is an emerging field of study; the following chapter will not only provide an overview, it will also delve deep into the variegated topics related to it. This chapter will elaborate the various aspects of this field in a lucid manner.

Sustainable Agriculture

Sustainable agriculture is farming in sustainable ways based on an understanding of ecosystem services, the study of relationships between organisms and their environment. It has been defined as "an integrated system of plant and animal production practices having a site-specific application that will last over the long term", for example:

- Satisfy human food and fiber needs

- Enhance environmental quality and the natural resource base upon which the agricultural economy depends

- Make the most efficient use of non-renewable resources and on-farm resources and integrate, where appropriate, natural biological cycles and controls

- Sustain the economic viability of farm operations

- Enhance the quality of life for farmers and society as a whole

History of the Term

The phrase was reportedly coined by the Australian agricultural scientist Gordon McClymont. Wes Jackson is credited with the first publication of the expression in his 1980 book *New Roots for Agriculture*. The term became popularly used in the late 1980s.

Farming and Natural Resources

Sustainable agriculture can be understood as an ecosystem approach to agriculture. Practices that can cause long-term damage to soil include excessive tilling of the soil (leading to erosion) and irrigation without adequate drainage (leading to salinization). Long-term experiments have provided some of the best data on how various practices affect soil properties essential to sustainability. In the United States a federal agency, USDA-Natural Resources Conservation Service, specializes in providing technical and financial assistance for those interested in pursuing natural resource conservation and production agriculture as compatible goals.

The most important factors for an individual site are sun, air, soil, nutrients, and water. Of the five, water and soil quality and quantity are most amenable to human intervention through time and labor.

Traditional farming methods had zero carbon footprint.

Although air and sunlight are available everywhere on Earth, crops also depend on soil nutrients and the availability of water. When farmers grow and harvest crops, they remove some of these nutrients from the soil. Without replenishment, land suffers from nutrient depletion and becomes either unusable or suffers from reduced yields. Sustainable agriculture depends on replenishing the soil while minimizing the use or need of non-renewable resources, such as natural gas (used in converting atmospheric nitrogen into synthetic fertilizer), or mineral ores (e.g., phosphate). Possible sources of nitrogen that would, in principle, be available indefinitely, include:

1. recycling crop waste and livestock or treated human manure

2. growing legume crops and forages such as peanuts or alfalfa that form symbioses with nitrogen-fixing bacteria called rhizobia

3. industrial production of nitrogen by the Haber process uses hydrogen, which is currently derived from natural gas (but this hydrogen could instead be made by electrolysis of water using electricity (perhaps from solar cells or windmills)) or

4. genetically engineering (non-legume) crops to form nitrogen-fixing symbioses or fix nitrogen without microbial symbionts.

The last option was proposed in the 1970s, but is only recently becoming feasible. Sustainable options for replacing other nutrient inputs (phosphorus, potassium, etc.) are more limited.

More realistic, and often overlooked, options include long-term crop rotations, returning to natural cycles that annually flood cultivated lands (returning lost nutrients indefinitely) such as the flooding of the Nile, the long-term use of biochar, and use of crop and livestock landraces that are adapted to less than ideal conditions such as pests, drought, or lack of nutrients.

Crops that require high levels of soil nutrients can be cultivated in a more sustainable manner if certain fertilizer management practices are adhered to.

Nationwide food producers require vast amounts of land and soil to produce food at an accelerated rate. This diminishes the nutrients in the soil and decimates the idea of sustainable agriculture, which is best built through local, regional agricultural methods.

Water

In some areas sufficient rainfall is available for crop growth, but many other areas require irrigation. For irrigation systems to be sustainable, they require proper management (to avoid salinization) and must not use more water from their source than is naturally replenishable. Otherwise, the water source effectively becomes a non-renewable resource. Improvements in water well drilling technology and submersible pumps, combined with the development of drip irrigation and low-pressure pivots, have made it possible to regularly achieve high crop yields in areas where reliance on rainfall alone had previously made successful agriculture unpredictable. However, this progress has come at a price. In many areas, such as the Ogallala Aquifer, the water is being used faster than it can be replenished.

Several steps must be taken to develop drought-resistant farming systems even in "normal" years with average rainfall. These measures include both policy and management actions:

1. improving water conservation and storage measures,
2. providing incentives for selection of drought-tolerant crop species,
3. using reduced-volume irrigation systems,
4. managing crops to reduce water loss, and
5. not planting crops at all.

Indicators for sustainable water resource development are:

- Internal renewable water resources. This is the average annual flow of rivers and groundwater generated from endogenous precipitation, after ensuring that there is no double counting. It represents the maximum amount of water resource produced within the boundaries of a country. This value, which is expressed as an average on a yearly basis, is invariant in time (except in the case of proved climate change). The indicator can be expressed in three different units: in absolute terms (km^3/yr), in mm/yr (it is a measure of the humidity of the country), and as a function of population (m^3/person per year).

- Global renewable water resources. This is the sum of internal renewable water resources and incoming flow originating outside the country. Unlike internal resources, this value can vary with time if upstream development reduces water availability at the border. Treaties ensuring a specific flow to be reserved from upstream to downstream countries may be taken into account in the computation of global water resources in both countries.

- Dependency ratio. This is the proportion of the global renewable water resources originating outside the country, expressed in percentage. It is an expression of the level to which the water resources of a country depend on neighbouring countries.

- Water withdrawal. In view of the limitations described above, only gross water withdrawal can be computed systematically on a country basis as a measure of water use. Absolute or per-person value of yearly water withdrawal gives a measure of the importance of water in the country's economy. When expressed in percentage of water resources, it shows the degree of pressure on water resources. A rough estimate shows that if water withdrawal exceeds a quarter of global renewable water resources of a country, water can be considered a limiting factor to development and, reciprocally, the pressure on water resources can affect all sectors, from agriculture to environment and fisheries.

Soil

Walls built to avoid water run-off

Soil erosion is fast becoming one of the world's severe problems. It is estimated that "more than a thousand million tonnes of southern Africa's soil are eroded every year. Experts predict that crop yields will be halved within thirty to fifty years if erosion continues at present rates." Soil erosion is not unique to Africa but is occurring worldwide. The phenomenon is being called *peak soil* as present large-scale factory farming techniques are jeopardizing humanity's ability to grow food in the present and in the future. Without efforts to improve soil management practices, the availability of arable soil will become increasingly problematic.

Some soil management techniques

- No-till farming

- Keyline design

- Growing windbreaks to hold the soil

- Incorporating organic matter back into fields

- Stop using chemical fertilizers (which contain salt)
- Protecting soil from water run-off (soil erosion)

Phosphate

Phosphate is a primary component in the chemical fertilizer which is applied in modern agricultural production. However, scientists estimate that rock phosphate reserves will be depleted in 50–100 years and that peak phosphorus will occur in about 2030. The phenomenon of peak phosphorus is expected to increase food prices as fertilizer costs increase as rock phosphate reserves become more difficult to extract. In the long term, phosphate will therefore have to be recovered and recycled from human and animal waste in order to maintain food production.

Land

As the global population increases and demand for food increases, there is pressure on land resources. Land can also be considered a finite resource on Earth. Expansion of agricultural land decreases biodiversity and contributes to deforestation. The Food and Agriculture Organisation of the United Nations estimates that in coming decades, cropland will continue to be lost to industrial and urban development, along with reclamation of wetlands, and conversion of forest to cultivation, resulting in the loss of biodiversity and increased soil erosion.

Energy for Agriculture

Energy is used all the way down the food chain from farm to fork. In industrial agriculture, energy is used in on-farm mechanisation, food processing, storage, and transportation processes. It has therefore been found that energy prices are closely linked to food prices. Oil is also used as an input in agricultural chemicals. Higher prices of non-renewable energy resources are projected by the International Energy Agency. Increased energy prices as a result of fossil fuel resources being depleted may therefore decrease global food security unless action is taken to 'decouple' fossil fuel energy from food production, with a move towards 'energy-smart' agricultural systems. The use of solar powered irrigation in Pakistan has come to be recognized as a leading example of energy use in creating a closed system for water irrigation in agricultural activity.

Economics

Socioeconomic aspects of sustainability are also partly understood. Regarding less concentrated farming, the best known analysis is Netting's study on smallholder systems through history. The Oxford Sustainable Group defines sustainability in this context in a much broader form, considering effect on all stakeholders in a 360 degree approach.

Given the finite supply of natural resources at any specific cost and location, agriculture that is inefficient or damaging to needed resources may eventually exhaust the available resources or the ability to afford and acquire them. It may also generate negative externality, such as pollution as well as financial and production costs. There are several studies incooperating these negative externalities in an economic analysis concerning ecosystem services, biodiversity, land degradation and sustainable land management. These include the The Economics of Ecosystems and Biodiversity (TEEB) study led by Pavan Sukhdev and the Economics of Land Degradation Initiative which seeks to establish an economic cost benefit analysis on the practice of sustainable land management and sustainable agriculture.

The way that crops are sold must be accounted for in the sustainability equation. Food sold locally does not require additional energy for transportation (including consumers). Food sold at a remote location, whether at a farmers' market or the supermarket, incurs a different set of energy cost for materials, labour, and transport.

Pursuing sustainable agriculture results in many localized benefits. Having the opportunities to sell products directly to consumers, rather than at wholesale or commodity prices, allows farmers to bring in optimal profit.

Methods

Polyculture practices in Andhra Pradesh

What grows where and how it is grown are a matter of choice. Two of the many possible practices of sustainable agriculture are crop rotation and soil amendment, both designed to ensure that crops being cultivated can obtain the necessary nutrients for healthy growth. Soil amendments would include using locally available compost from community recycling centers. These community recycling centers help produce the compost needed by the local organic farms.

Using community recycling from yard and kitchen waste utilizes a local area's commonly available resources. These resources in the past were thrown away into large waste disposal sites, are now used to produce low cost organic compost for organic farming. Other practices includes growing a diverse number of perennial crops in a single field, each of which would grow in separate season so as not to compete with each other for natural resources. This system would result in increased resistance to diseases and decreased effects of erosion and loss of nutrients in soil. Nitrogen fixation from legumes, for example, used in conjunction with plants that rely on nitrate from soil for growth, helps to allow the land to be reused annually. Legumes will grow for a season and replenish the soil with ammonium and nitrate, and the next season other plants can be seeded and grown in the field in preparation for harvest.

Rotational grazing practices in use with paddocks

Monoculture, a method of growing only one crop at a time in a given field, is a very widespread practice, but there are questions about its sustainability, especially if the same crop is grown every year. Today it is realized to get around this problem local cities and farms can work together to produce the needed compost for the farmers around them.

This combined with growing a mixture of crops (polyculture) sometimes reduces disease or pest problems but polyculture has rarely, if ever, been compared to the more widespread practice of growing different crops in successive years (crop rotation) with the same overall crop diversity. Cropping systems that include a variety of crops (polyculture and/or rotation) may also replenish nitrogen (if legumes are included) and may also use resources such as sunlight, water, or nutrients more efficiently (Field Crops Res. 34:239).

Replacing a natural ecosystem with a few specifically chosen plant varieties reduces the genetic diversity found in wildlife and makes the organisms susceptible to widespread disease. The Great Irish Famine (1845–1849) is a well-known example of the dangers of monoculture. In practice, there is no single approach to sustainable agriculture, as the precise goals and methods must be adapted to each individual case. There may be some techniques of farming that are inherently in conflict with the concept of sustainability, but there is widespread misunderstanding on effects of some practices.

Today the growth of local farmers' markets offer small farms the ability to sell the products that they have grown back to the cities that they got the recycled compost from. By using local recycling this will help move people away from the slash-and-burn techniques that are the characteristic feature of shifting cultivators are often cited as inherently destructive, yet slash-and-burn cultivation has been practiced in the Amazon for at least 6000 years; serious deforestation did not begin until the 1970s, largely as the result of Brazilian government programs and policies.

To note that it may not have been slash-and-burn so much as slash-and-char, which with the addition of organic matter produces terra preta, one of the richest soils on Earth and the only one that regenerates itself.

There are also many ways to practice sustainable animal husbandry. Some of the key tools to grazing management include fencing off the grazing area into smaller areas called paddocks, lowering stock density, and moving the stock between paddocks frequently.

Sustainable Intensification

In light of concerns about food security, human population growth and dwindling land suitable for agriculture, sustainable intensive farming practises are needed to maintain high crop yields, while maintaining soil health and ecosystem services. The capacity for ecosystem services to be strong enough to allow a reduction in use of synthetic, non renewable inputs whilst maintaining or even boosting yields has been the subject of much debate. Recent work in the globally important irrigated rice production system of east Asia has suggested that - in relation to pest management at least - promoting the ecosystem service of biological control using nectar plants can reduce the need for insecticides by 70% whilst delivering a 5% yield advantage compared with standard practice.

Soil Treatment

Sheet steaming with a MSD/moeschle steam boiler (left side)

Soil steaming can be used as an ecological alternative to chemicals for soil sterilization. Different methods are available to induce steam into the soil in order to kill pests and increase soil health.

Solarizing is based on the same principle, used to increase the temperature of the soil to kill pathogens and pests.

Certain crops act as natural biofumigants, releasing pest suppressing compounds. Mustard, radishes, and other plants in the brassica family are best known for this effect. There exist varieties of mustard shown to be almost as effective as synthetic fumigants at a similar or lesser cost.

Off-farm Impacts

A farm that is able to "produce perpetually", yet has negative effects on environmental quality elsewhere is not sustainable agriculture. An example of a case in which a global view may be warranted is over-application of synthetic fertilizer or animal manures, which can improve productivity of a farm but can pollute nearby rivers and coastal waters (eutrophication). The other extreme can also be undesirable, as the problem of low crop yields due to exhaustion of nutrients in the soil has been related to rainforest destruction, as in the case of slash and burn farming for livestock feed.In Asia, specific land for sustainable farming is about 12.5 acres which includes land for animal fodder, cereals productions lands for some cash crops and even recycling of related food crops.In some cases even a small unit of aquaculture is also included in this number (AARI-1996)

Sustainability affects overall production, which must increase to meet the increasing food and fiber requirements as the world's human population expands to a projected 9.3 billion people by 2050. Increased production may come from creating new farmland, which may ameliorate carbon dioxide emissions if done through reclamation of desert as in Israel and Palestine, or may worsen emissions if done through slash and burn farming, as in Brazil.

International Policy

Sustainable agriculture has become a topic of interest in the international policy arena, especially with regards to its potential to reduce the risks associated with a changing climate and growing human population.

The Commission on Sustainable Agriculture and Climate Change, as part of its recommendations for policy makers on achieving food security in the face of climate change, urged that sustainable agriculture must be integrated into national and international policy. The Commission stressed that increasing weather variability and climate shocks will negatively affect agricultural yields, necessitating early action to drive change in agricultural production systems towards increasing resilience. It also called for dramatically increased investments in sustainable agriculture in the next decade, including in national research and development budgets, land rehabilitation, economic incentives, and infrastructure improvement.

Urban Planning

There has been considerable debate about which form of human residential habitat may be a better social form for sustainable agriculture.

Many environmentalists advocate urban developments with high population density as a way of preserving agricultural land and maximizing energy efficiency. However, others have theorized that sustainable ecocities, or ecovillages which combine habitation and farming with close proximity between producers and consumers, may provide greater sustainability.

The use of available city space (e.g., rooftop gardens, community gardens, garden sharing, and other forms of urban agriculture) for cooperative food production is another way to achieve greater sustainability.

One of the latest ideas in achieving sustainable agriculture involves shifting the production of food plants from major factory farming operations to large, urban, technical facilities called vertical farms. The advantages of vertical farming include year-round production, isolation from pests and diseases, controllable resource recycling, and on-site production that reduces transportation costs. While a vertical farm has yet to become a reality, the idea is gaining momentum among those who believe that current sustainable farming methods will be insufficient to provide for a growing global population.

Criticism

Efforts toward more sustainable agriculture are supported in the sustainability community, however, these are often viewed only as incremental steps and not as an end. Some foresee a true sustainable steady state economy that may be very different from today's: greatly reduced energy usage, minimal ecological footprint, fewer consumer packaged goods, local purchasing with short food supply chains, little processed foods, more home and community gardens, etc. Agriculture would be very different in this type of sustainable economy.

Form of Sustainable Agriculture

Sustainable Planting

Sustainable planting is an approach to planting design and landscaping-gardening that balances the need for resource conservation with the needs of farmers pursuing their livelihood. The demand on resources, specifically land/crops, is constantly increasing due to the long human lifespan. It is a form of sustainable agriculture and, "it considers long-term as well as short-term economics because sustainability is readily defined as forever, that is, agricultural environments that are designed to promote endless regeneration".

The idea of sustainable planting can be dated back millennia, when the ancient Greeks and Chinese practised organic farming, the oldest method of farming. Later this practice was largely replaced by inorganic farming. In 1907 Franklin H. King in his book ``*Farmers of Forty Centuries*`` discussed the advantages of sustainable agriculture, and warned that sustainable practices would be vital to farming in the future.

Advantages of Sustainable Planting

TABLE 1.1. Sustainability, External Inputs Needed, and Labor Requirements of Selected Plant Disease Management Practices of Traditional Farmers.

Practice	Sustainable?	External Inputs	Labor
adjusting crop density	Yes	Low	Low
adjusting depth of planting	Yes	Low	Low
adjusting time of planting	Yes	Low	Low
altering of plant and crop architecture	Yes	Low	High
biological control (soilborne pathogens)	Yes	High	High
burning	Yes a	Low	High
fallowing	Yes	Low	Low
flooding	Yes	Low	High
manipulating shade	Yes	Low	Low
mulching	Yes	High	High
multistory cropping	Yes	Low	Low
multiple cropping	Yes	Low	High
planting diverse crops	Yes	Low	Low
planting in raised beds	Yes	High	High
rotation	Yes	Low	Low
site selection	Yes	Low	Low
tillage	No	Low	High
using organic amendments	Yes	High	High
weed control	No	Low	High

^aUnder high population pressure the slash and burn system is neither stable nor sustainable.

Sustainability, external inputs needed, and labour requirements of selected plant disease management practices of traditional farmers.

Sustainable planting does not necessarily consist only of planting native species. Some ecosystems may benefit from any increase in biomass, from the introduction of certain non-native species, or any increase in biodiversity. In the case of disturbed areas, such as areas where energy pipelines have been installed or areas where military activity has taken place, some exotic/non-native plants may fare better than the displaced, native inhabitants, in the process increasing the biodiversity and biological biomass. Sustainable planting may also involve crop rotation provided that they are used effectively. At the very least, constant crop rotation will prevent soil erosion, by protecting topsoil from wind and draining water. Effective crop rotation allows enough time for pest pressure on crops to be significantly reduced, and for soil nutrients to be replenished. This, in turn, reduces the need for chemical fertilizers and pesticides. Specifically in terms of industrial agriculture, increasing the genetic diversity of crops by introducing new germplasm can significantly impact the heartiness of crops (cost permitting).

Methods of Planting Sustainably

Entomopathogenic Nematodes Application

Entomopathogenic nematode are parasites that are beneficial because they can be used to control pests in agriculture or forestry. This serves to drastically reduce the need for using pesticides to get rid of pests in planting. Furthermore EPN(Entomopathogenic Nematodes) can also serve as excellent resources for understanding biological, ecological and evolutionary processes involving other soil organisms. EPN form a stress–resistant stage known as the infective juvenile (IJ). The infective juvenile spread in the soil and infect suitable insect hosts. Upon entering the insect they move to the hemolymph where they recover from their stagnated state of development and release their bacterial symbionts. The bacterial symbionts reproduce and release toxins, which then kill the host insect. Compared to pesticides as they have very broad host range and less likely to induce insect form immunity to them. Unlike many other chemical insecticides, they do not poison non-target animals.

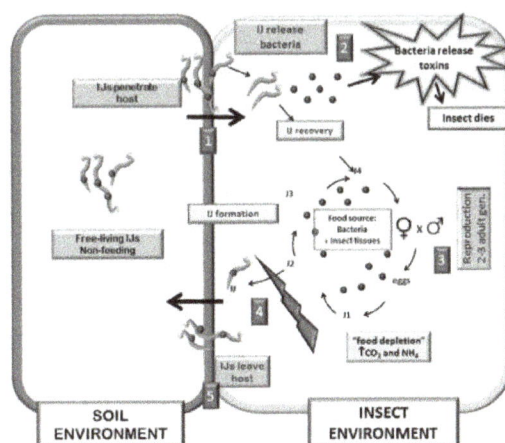

Fig. 1.2 Generalized life cycle of entomopathogenic nematodes and their bacterial symbionts

Generalized life cycle of entomopathogenic nematodes and their bacterial symbionts.

An example of a property using the xeriscaping method of sustainable planting. This method reduces the need for water which is often in limited supply in arid regions.

Limitations of EPN

Entomopathogenic Nematodes have a very limited shelf life.Furthermore, they are negatively affected by high temperature and dry conditions as they are organisms mostly suited to water. When transported, they may get exposed to these unfavourable conditions, resulting in shortening their shelf life and viability. The type of soil they are applied to may also limit their effectiveness, as most of them might die before finding a suitable host. More research is being done to discover ways to overcome these limitations.

Improving Water Efficiency

Water efficiency can be improved by reducing the need for irrigation and using alternative methods. Such methods includes: researching on drought resistant crops, monitoring plant transpiration and reducing soil evaporation.

Soil Water Evaporation

It has been discovered that abstinence from soil tillage before planting and leaving the plant residue after harvesting reduces soil water evaporation; It also serves to effectively prevent soil erosion.

Crop residues reduce the evaporation of water from soil by covering the surface of the soil, this results in a lower surface soil temperature and reduction of wind effects. No tilling reduces the need for irrigation, using this method helps with the efficient use of water. Using the Xeriscaping approach is another possible way to conserve water.

Drought Resistant Crops

Drought resistant crops have been researched extensively as a means to overcome the issue of water shortage. They are modified genetically so they can adapt in an environment with little water. This is beneficial as it reduces the need for irrigation and helps conserve water. Although they have been extensively researched, significant results have not been achieved as most of the successful species will have no overall impact on water conservation. However, some grains like rice, for example, have been successfully genetically modified to be drought resistant reducing the farmers need for irrigation.

Native Plants Versus Exotic Plants

There is some debate among researchers and sustainability advocates whether it is more sustainable to cultivate plants that are native to a bioregion or choose plants based on the needs of the community or environment (regardless of the plants natural bioregion). Inconsistent, scientific definitions of words like native and sustainable are largely responsible for the controversy surrounding this issue. Because evolution of a plant species can be correlated with natural migration of the species to new areas, it is unclear whether or not this would classify individuals in this new area as native or exotic. With current technology and scientific understanding, the fossil record does not allow for accurate tracking of plant evolution. There is no standard length of time a species must inhabit an area before it becomes a native species to that area. Species may be considered exotic or invasive if they have been introduced by humans and are detrimental to other species in the area. Some sustainability advocates champion for the use of native plants when water shortage and soil nutrients are an issue because some exotic plants require different amounts of water and nutrients than what will occur naturally in an area. Conversely, the use of exotic plants can be argued for when the given species will add nutrients to the soil, like during crop rotation, or will serve another purpose that native plants could not. There is no blanket preference for native species or exotic plants as the benefits are very situation-specific.

Practical Examples

- When creating new roads or widening current roads, the Nevada Department of Transportation will reserve topsoil and native plants for donation. These materials can be used by home gardeners or for large-scale agricultural operations to supplement or replace imported materials. This decreases the need to import soil and plants that may disturb the ecosystem. Native plants and soil may also be better suited to local weather conditions and require less water and fewer soil additives (fertilizers).

- The National Wildlife Federation has created a network of gardens in the USA that are Certified Wildlife Habitats through the Garden for Wildlife Program that started in 1973.

- The Grain for Green Program in China that pays farmers to convert their retired farmland back into forests or other natural landscapes (usually what it was before farming).

- Over one-hundred peer-reviewed studies published during or after the year 2010 show that urban green space lowers air and ground temperatures and improves air quality. The size, quality and density of urban green spaces are positively correlated with the derived benefits. Sustainable urban green spaces dramatically reduce heat-stress related deaths and hospitalizations in urban communities and improve the overall health of individuals within those communities. Some cities trying to increase urban green space include Ottawa, Toronto, Winnipeg, Calgary, Whistler, Chequamegon, and Dauphin Island.

References

- Wes Jackson, New Roots for Agriculture. Foreword by Wendell Berry. University of Nebraska Press. ISBN 0803275625

- Kunstler, James Howard (2012). Too Much Magic; Wishful Thinking, Technology, and the Fate of the Nation. Atlantic Monthly Press. ISBN 978-0-8021-9438-1.

- McKibben, D, ed. (2010). The Post Carbon Reader: Managing the 21st Centery Sustainability Crisis. Watershed Media. ISBN 978-0-9709500-6-2.

- Tomich, Tom (2016). Sustainable Agriculture Research and Education Program (PDF). Davis, California: University of California.

- Flint, R. Warren (2012). Practice of Sustainable Community Development A Participatory Framework for Change. Dordrecht : Springer. ISBN 1-4614-5099-3.

- Thurston, David H. (1991). Sustainable Practices for Plant Disease Management in Traditional Farmer Systems. HarperCollins Canada / Westview S/Dis. ISBN 978-0813383637.

- Webb, Robert H. (2009). The Mojave Desert : ecosystem processes and sustainability. University of Nevada. ISBN 9780874177763.

- Campos, Herrera R. (2015). Nematode Pathogenesis of insects and other pests. (1 ed.). Springer. pp. 4–6. ISBN 978-3-319-18266-7. Retrieved 3 February 2016.

- Buckstrup, Michelle (1997). "Native vs. Exotic for the Home Landscape". Cornell University Department of Horticulture. Cornell University. Retrieved 2016. Check date values in: |access-date= (help)

- Yuan, Zhen (2014). China's Grain for Green Program A Review of the Largest Ecological Restoration and Rural Development Program in the World. Cham : Springer International Publishing. ISBN 3-319-11504-9.

Basic Concepts of Sustainable Agriculture

This chapter carefully elaborates the basic concepts of sustainable agriculture to provide a complete understanding of sustainable agriculture. Concepts like crop rotation, monoculture, polyculture, cover crop, weed control among others are described in the following chapter in a critical and systematic manner.

Crop Rotation

Satellite image of circular crop fields in Kansas in late June 2001. Healthy, growing crops are green. Corn would be growing into leafy stalks by then. Sorghum, which resembles corn, grows more slowly and would be much smaller and therefore, (possibly) paler. Wheat is a brilliant yellow as harvest occurs in June. Fields of brown have been recently harvested and plowed under or lie fallow for the year.

Effects of crop rotation and monoculture at the Swojec Experimental Farm, Wroclaw University of Environmental and Life Sciences. In the front field, the "Norfolk" crop rotation sequence (potatoes, oats, peas, rye) is being applied; in the back field, rye has been grown for 45 years in a row.

Crop rotation is the practice of growing a series of dissimilar or different types of crops in the same area in sequenced seasons. It is done so that the soil of farms is not used to only one type of nutrient. It helps in reducing soil erosion and increases soil fertility and crop yield.

Growing the same crop in the same place for many years in a row disproportionately depletes the soil of certain nutrients. With rotation, a crop that leaches the soil of one kind of nutrient is followed during the next growing season by a dissimilar crop that returns that nutrient to the soil or draws a different ratio of nutrients. In addition, crop rotation mitigates the buildup of pathogens and pests that often occurs when one species is continuously cropped, and can also improve soil structure and fertility by increasing biomass from varied root structures.

Crop rotation is used in both conventional and organic farming systems.

History

It has long been recognized that suitable rotations – such as planting spring crops for livestock in place of grains for human consumption – make it possible to restore or to maintain a productive soil. Middle Eastern farmers practiced crop rotation in 6000 BC without understanding the chemistry, alternately planting legumes and cereals. In the Bible chapter of Leviticus 25, God instructs the Israelites to observe a 'Sabbath of the Land'. Every seventh year they would not till, prune or even control insects. The Roman writer, Cato the Elder, recommended that farmers "save carefully goat, sheep, cattle, and all other dung". In Europe, since the times of Charlemagne, there was a transition from a two-field crop rotation to a three-field crop rotation. Under a two-field rotation, half the land was planted in a year, while the other half lay fallow. Then, in the next year, the two fields were reversed.

From the end of the Middle Ages until the 20th century, three-year rotation was practiced by farmers in Europe. Under three-field rotation, the land was divided into three parts. One section was planted in the autumn with rye or winter wheat, followed by spring oats or barley, or other crops such as peas, lentils, or beans and the third field was left fallow. The three fields were rotated in this manner so that every three years, a field would rest and be fallow. Under the two-field system, if one has a total of 600 acres (2.4 km²) of fertile land, one would only plant 300 acres. Under the new three-field rotation system, one would plant (and therefore harvest) 400 acres. But, the additional crops had a more significant effect than mere productivity. Since the spring crops were mostly legumes, they increased the overall nutrition of the people of Northern Europe.

A four-field rotation was pioneered by farmers, namely in the region Waasland in the early 16th century and popularised by the British agriculturist Charles Townshend in the 18th century. The system (wheat, turnips, barley and clover), opened up a fodder crop and grazing crop allowing livestock to be bred year-round. The four-field crop rotation was a key development in the British Agricultural Revolution.

George Washington Carver studied crop rotation methods in the United States, teaching southern farmers to rotate soil-depleting crops like cotton with soil-enriching crops like peanuts and peas.

In the Green Revolution, the traditional practice of crop rotation gave way in some parts of the world to the practice of supplementing the chemical inputs to the soil through top dressing with fertilizers, e.g. adding ammonium nitrate or urea and restoring soil pH with lime in the search for increased yields, preparing soil for specialist crops, and seeking to reduce waste and inefficiency by simplifying planting and harvesting.

Crop Choice

A preliminary assessment of crop interrelationships can be found in how each crop: (1) contributes to soil organic matter (SOM) content, (2) provides for pest management, (3) manages deficient or excess nutrients, and (4) how it contributes to or controls for soil erosion.

Crop choice is often a related to the goal the farmer is looking to achieve with the rotation, which could be weed management, increasing available nitrogen in the soil, controlling for erosion, or increasing soil structure and biomass, to name a few. When discussing crop rotations, crops are classified in different ways depending on what quality is being assessed: by family, by nutrient needs/benefits, and/or by profitability (i.e. cash crop versus cover crop). For example, giving adequate attention to plant family is essential to mitigating pests and pathogens. However, many farmers have success managing rotations by planning sequencing and cover crops around desirable cash crops. The following is a simplified classification based on crop quality and purpose.

Row Crops

Many crops which are critical for the market, like vegetables, are row crops (that is, grown in tight rows). While often the most profitable for farmers, these crops are more taxing on the soil Row crops typically have low biomass and shallow roots: this means the plant contributes low residue to the surrounding soil and has limited effects on structure. With much of the soil around the plant is exposed to disruption by rainfall and traffic, fields with row crops experience faster break down of organic matter by microbes, leaving fewer nutrients for future plants.

In short, while these crops may be profitable for the farm, they are nutrient depleting. Crop rotation practices exist to strike a balance between short-term profitability and long-term productivity.

Legumes

A great advantage of crop rotation comes from the interrelationship of nitrogen fixing-crops with nitrogen demanding crops. Legumes, like alfalfa and clover, collect available nitrogen from the soil in nodules on their root structure. When the plant is harvested, the biomass of uncollected roots breaks down, making the stored nitrogen available to future crops. Legumes are also a valued green manure: a crop that collects nutrients and fixes them at soil depths accessible to future crops.

In addition, legumes have heavy tap roots that burrow deep into the ground, lifting soil for better tilth and absorption of water.

Grasses and Cereals

Cereal and grasses are frequent cover crops because of the many advantages they supply to soil quality and structure. The dense and far-reaching root systems give ample structure to surrounding soil and provide significant biomass for soil organic matter.

Grasses and cereals are key in weed management as they compete with undesired plants for soil space and nutrients.

Green Manure

Green manure is a crop that is mixed into the soil. Both nitrogen-fixing legumes and nutrient scavengers, like grasses, can be used as green manure. Green manure of legumes is an excellent source of nitrogen, especially for organic systems, however, legume biomass doesn't contribute to lasting soil organic matter like grasses do.

Planning a Rotation

There are numerous factors that must be taken into consideration when planning a crop rotation. Planning an effective rotation requires weighing fixed and fluctuating production circumstances, including, but not limited to: market, farm size, labor supply, climate, soil type, growing practices, etc. Moreover, a crop rotation must consider in what condition one crop will leave the soil for the succeeding crop and how one crop can be seeded with another crop. For example, a nitrogen-fixing crop, like a legume, should always proceed a nitrogen depleting one; similarly, a low residue crop (i.e. a crop with low biomass) should be offset with a high biomass cover crop, like a mixture of grasses and legumes.

There is no limit to the number of crops that can be used in a rotation, or the amount of time a rotation takes to complete. Decisions about rotations are made years prior, seasons prior, or even at the very last minute when an opportunity to increase profits or soil quality presents itself. In short, there is no singular formula for rotation, but many considerations to take into account.

Implementation

Crop rotation systems may be enriched by the influences of other practices such as the addition of livestock and manure, intercropping or multiple cropping, and organic management low in pesticides and synthetic fertilizers.

Incorporation of Livestock

Introducing livestock makes the most efficient use of critical sod and cover crops; livestock (through manure) are able to distribute the nutrients in these crops throughout the soil rather than removing nutrients from the farm through the sale of hay. In systems where use of farm livestock would violate reservations growers or consumers may have about animal exploitation, efforts are made to surrogate this input through livestock in the soil, namely worms and microorganisms.

In Sub-Saharan Africa, as animal husbandry becomes less of a nomadic practice many herders have begun integrating crop production into their practice. This is known as mixed farming, or the practice of crop cultivation with the incorporation of raising cattle, sheep and/or goats by the same economic entity, is increasingly common. This interaction between the animal, the land and the crops are being done on a small scale all across this region. Crop residues provide animal feed, while the animals provide manure for replenishing crop nutrients and draft power. Both processes are extremely important in this region of the world as it is expensive and logistically unfeasible to transport in synthetic fertilizers and large-scale machinery. As an additional benefit, the cattle, sheep and/or goat provide milk and can act as a cash crop in the times of economic hardship.

Organic Farming

Crop rotation is a required practice in order for a farm to receive organic certification in the United States. The "Crop Rotation Practice Standard" for the National Organic Program under the U.S. Code of Federal Regulations, section §205.205, states that:

Farmers are required to implement a crop rotation that maintains or builds soil organic matter, works to control pests, manages and conserves nutrients, and protects against erosion. Producers of perennial crops that aren't rotated may utilize other practices, such as cover crops, to maintain soil health.

In addition to lowering the need for inputs by controlling for pests and weeds and increasing available nutrients, crop rotation helps organic growers increase the amount of biodiversity on their farms. Biodiversity is also a requirement of organic certification, however, there are no rules in place to regulate or reinforce this standard. Increasing the biodiversity of crops has beneficial effects on the surrounding ecosystem and can host a greater diversity of fauna, insects, and beneficial microorganism in the soil.< Some studies point to increased nutrient availability from crop rotation under organic systems compared to conventional practices as organic practices are less likely to inhibit of beneficial microbes in soil organic matter.

While multiple cropping and intercropping benefit from many of the same principals as crop rotation, they do not satisfy the requirement under the NOP.

Intercropping

Multiple cropping systems, such as intercropping or companion planting, offer more diversity and complexity within the same season or rotation, for example the three sisters. An example of companion planting is the inter-planting of corn with pole beans and vining squash or pumpkins. In this system, the beans provide nitrogen; the corn provides support for the beans and a "screen" against squash vine borer; the vining squash provides a weed suppressive canopy and discourages corn-hungry raccoons.

Double-cropping is common where two crops, typically of different species, are grown sequentially in the same growing season, or where one crop (e.g. vegetable) is grown continuously with a cover crop (e.g. wheat). This is advantageous for small farms, who often cannot afford to leave cover crops to replenish the soil for extended periods of time, as larger farms can. When multiple cropping is implemented on small farms, these systems can maximize benefits of crop rotation on available land resources.

Benefits

Agronomists describe the benefits to yield in rotated crops as "The Rotation Effect". There are many found benefits of rotation systems: however, there is no specific scientific basis for the sometimes 10-25% yield increase in a crop grown in rotation versus monoculture. The factors related to the increase are simply described as alleviation of the negative factors of monoculture cropping systems. Explanations due to improved nutrition; pest, pathogen, and weed stress reduction; and improved soil structure have been found in some cases to be correlated, but causation has not been determined for the majority of cropping systems.

Other benefits of rotation cropping systems include production cost advantages. Overall financial risks are more widely distributed over more diverse production of crops and/or livestock. Less reliance is placed on purchased inputs and over time crops can maintain production goals with fewer inputs. This in tandem with greater short and long term yields makes rotation a powerful tool for improving agricultural systems.

Soil Organic Matter

The use of different species in rotation allows for increased soil organic matter (SOM), greater soil structure, and improvement of the chemical and biological soil environment for crops. With more SOM, water infiltration and retention improves, providing increased drought tolerance and decreased erosion.

Soil organic matter is a mix of decaying material from biomass with active microorganisms. Crop rotation, by nature, increases exposure to biomass from sod, green manure, and a various other plant debris. The reduced need for intensive tillage under crop rotation allows biomass aggregation to lead to greater nutrient retention and utilization, decreasing the need for added nutrients. With tillage, disruption and oxidation of soil creates a less conducive environment for diversity and proliferation of microorganisms in the soil. These microorganisms are what make nutrients available to plants. So, where "active" soil organic matter is a key to productive soil, soil with low microbial activity provides significantly fewer nutrients to plants; this is true even though the quantity of biomass left in the soil may be the same.

Soil microorganisms also decrease pathogen and pest activity through competition. In addition, plants produce root exudates and other chemicals which manipulate their soil environment as well as their weed environment. Thus rotation allows increased yields from nutrient availability but also alleviation of allelopathy and competitive weed environments.

Carbon Sequestration

Studies have shown that crop rotations greatly increase soil organic carbon (SOC) content, the main constituent of soil organic matter. Carbon, along with hydrogen and oxygen, is a macro-nutrient for plants. Highly diverse rotations spanning long periods of time have shown to be even more effective in increasing SOC, while soil disturbances (e.g. from tillage) are responsible for exponential decline in SOC levels. In Brazil, conservation to no-till methods combined with intensive crop rotations has been shown an SOC sequestration rate of 0.41 tonnes per hectare per year.

In addition to enhancing crop productivity, sequestration of atmospheric carbon has great implications in reducing rates of climate change by removing carbon dioxide from the air.

Nitrogen Fixing

Rotating crops adds nutrients to the soil. Legumes, plants of the family Fabaceae, for instance, have nodules on their roots which contain nitrogen-fixing bacteria called rhizobia. It therefore makes good sense agriculturally to alternate them with cereals (family Poaceae) and other plants that require nitrates.

Pathogen and Pest Control

Crop rotation is also used to control pests and diseases that can become established in the soil over time. The changing of crops in a sequence decreases the population level of pests by (1) interrupting pest life cycles and (2) interrupting pest habitat. Plants within the same taxonomic family tend to have similar pests and pathogens. By regularly changing crops and keeping the soil occupied by cover crops instead of lying fallow, pest cycles can be broken or limited, especially cycles that benefit from overwintering in residue. For example, root-knot nematode is a serious problem for some plants in warm climates and sandy soils, where it slowly builds up to high levels in the soil, and can severely damage plant productivity by cutting off circulation from the plant roots. Growing a crop that is not a host for root-knot nematode for one season greatly reduces the level of the nematode in the soil, thus making it possible to grow a susceptible crop the following season without needing soil fumigation.

This principle is of particular use in organic farming, where pest control must be achieved without synthetic pesticides.

Weed Management

Integrating certain crops, especially cover crops, into crop rotations is of particular value to weed management. These crops crowd out weed through competition. In addition, the sod and compost from cover crops and green manure slows the growth of what weeds are still able to make it through the soil, giving the crops further competitive advantage. By removing slowing the growth and proliferation of weeds while cover crops are cultivated, farmers greatly reduce the presence of weeds for future crops, including shallow rooted and row crops, which are less resistant to weeds. Cover crops are, therefore, considered conservation crops because they protect otherwise fallow land from becoming overrun with weeds.

This system has advantages over other common practices for weeds management, such as tillage. Tillage is meant to inhibit growth of weeds by overturning the soil; however, this has a countering effect of exposing weed seeds that may have gotten buried and burying valuable crop seeds. Under crop rotation, the number of viable seeds in the soil is reduced through the reduction of the weed population.

Preventing Soil Erosion

Crop rotation can significantly reduce the amount of soil lost from erosion by water. In areas that are highly susceptible to erosion, farm management practices such as zero and reduced tillage can be supplemented with specific crop rotation methods to reduce raindrop impact, sediment detachment, sediment transport, surface runoff, and soil loss.

Protection against soil loss is maximized with rotation methods that leave the greatest mass of crop stubble (plant residue left after harvest) on top of the soil. Stubble cover in contact with the soil minimizes erosion from water by reducing overland flow velocity, stream power, and thus the ability of the water to detach and transport sediment. Soil Erosion and Cill prevent the disruption and detachment of soil aggregates that cause macropores to block, infiltration to decline, and runoff to increase. This significantly improves the resilience of soils when subjected to periods of erosion and stress.

The effect of crop rotation on erosion control varies by climate. In regions under relatively consistent climate conditions, where annual rainfall and temperature levels are assumed, rigid crop rotations can produce sufficient plant growth and soil cover. In regions where climate conditions are less predictable, and unexpected periods of rain and drought may occur, a more flexible approach for soil cover by crop rotation is necessary. An opportunity cropping system promotes adequate soil cover under these erratic climate conditions. In an opportunity cropping system, crops are grown when soil water is adequate and there is a reliable sowing window. This form of cropping system is likely to produce better soil cover than a rigid crop rotation because crops are only sown under optimal conditions, whereas rigid systems are not necessarily sown in the best conditions available.

Crop rotations also affect the timing and length of when a field is subject to fallow. This is very important because depending on a particular region's climate, a field could be the most vulnerable to erosion when it is under fallow. Efficient fallow management is an essential part of reducing erosion in a crop rotation system. Zero tillage is a fundamental management practice that promotes crop stubble retention under longer unplanned fallows when crops cannot be planted. Such management practices that succeed in retaining suitable soil cover in areas under fallow will ultimately reduce soil loss.

Biodiversity

Increasing the biodiversity of crops has beneficial effects on the surrounding ecosystem and can host a greater diversity of fauna, insects, and beneficial microorganisms in the soil. Some studies point to increased nutrient availability from crop rotation under organic systems compared to conventional practices as organic practices are less likely to inhibit of beneficial microbes in soil organic matter, such as arbuscular mycorrhizae, which increase nutrient uptake in plants. Increasing biodiversity also increases the resilience of agro-ecological systems.

Farm Productivity

Crop rotation contributes to increased yields through improved soil nutrition. By requiring planting and harvesting of different crops at different times, more land can be farmed with the same amount of machinery and labour.

Risk Management

Different crops in the rotation can reduce the risks of adverse weather for the individual farmer.

Challenges

While crop rotation requires a great deal of planning, crop choice must respond to a number of fixed conditions (soil type, topography, climate, and irrigation) in addition to conditions that may change dramatically from year to the next (weather, market, labor supply). In this way, it is unwise to plan to crops years in advance. Improper implementation of a crop rotation plan may lead to imbalances in the soil nutrient composition or a buildup of pathogens affecting a critical crop. The consequences of faulty rotation may take years to become apparent even to experienced soil scientists and can take just as long to correct.

Many challenges exist within the practices associated with crop rotation. For example, green manure from legumes can lead to an invasion of snails or slugs and the decay from green manure can occasionally suppress the growth of other crops.

Soil Conditioner

A soil conditioner is a product which is added to soil to improve the soil's physical qualities, especially its ability to provide nutrition for plants. In general usage, the term "soil conditioner" is often thought of as a subset of the category soil amendments, which more often is understood to include a wide range of fertilizers and non-organic materials.

Soil conditioners can be used to improve poor soils, or to rebuild soils which have been damaged by improper management. They can make poor soils more usable, and can be used to maintain soils in peak condition.

Composition

A wide variety of materials have been described as soil conditioners due to their ability to improve soil quality. Some examples include biochar, bone meal, blood meal, coffee grounds, compost, compost tea, coir, manure, straw, peat, sphagnum moss, vermiculite, sulfur, lime, hydroabsorbant polymers, and biosolids.

Many soil conditioners come in the form of certified organic products, for people concerned with maintaining organic crops or organic gardens. Soil conditioners of almost every description are readily available from online stores or local nurseries as well as garden supply stores.

Purpose

Soil Structure

The most common use of soil conditioners is to improve soil structure. Soils tend to become compacted over time. Soil compaction impedes root growth, decreasing the ability of plants to take up nutrients and water. Soil conditioners can add more loft and texture to keep the soil loose.

Soil Nutrients

For centuries people have been adding things to poor soils to improve their ability to support healthy plant growth. Some of these materials, such as compost, clay and peat, are still used extensively today. Many soil amendments also add nutrients such as carbon and nitrogen, as well as beneficial bacteria.

Additional nutrients, such as calcium, magnesium and phosphorus, may be augmented by amendments as well. This enriches the soil, allowing plants to grow bigger and stronger.

Cation Exchange

Soil amendments can also greatly increase the cation exchange capacity of soils. Soils act as the storehouses of plant nutrients. The relative ability of soils to store one particular group of nutrients, the cations, is referred to as cation exchange capacity or CEC. The most common soil cations are calcium, magnesium, potassium, ammonium, hydrogen, and sodium.

The total number of cations a soil can hold, its total negative charge, is the soil's cation exchange capacity. The higher the CEC, the higher the negative charge and the more cations that can be held and exchanged with plant roots, providing them with the nutrition they require.

Water Retention

Soil conditioners may be used to improve water retention in dry, coarse soils which are not holding water well. The addition of organic material for instance can greatly improve the water retention abilities of sandy soils and they can be added to adjust the pH of the soil to meet the needs of specific plants or to make highly acidic or alkaline soils more usable. The possibility of using other materials to assume the role of composts and clays in improving the soil was investigated on a scientific basis earlier in the 20th century, and the term soil conditioning was coined. The criteria by which such materials are judged most often remains their cost-effectiveness, their ability to increase soil moisture for longer periods, stimulate microbiological activity, increase nutrient levels and improve plant survival rates.

The first synthetic soil conditioners were introduced in the 1950s, when the chemical hydrolysed polyacrylonitrile was the most used. Because of their ability to absorb several hundred times their own weight in water, polyacrylamides and polymethacrylates (also known as hydroabsorbent polymers, superabsorbent polymers or hydrogels) were tested in agriculture, horticulture and landscaping beginning in the 1960s.

Interest disappeared when experiments proved them to be phytotoxic due to their high acrylamide monomer residue. Although manufacturing advances later brought the monomer concentration down below the toxic level, scientific literature shows few successes in utilizing these polymers for increasing plant quality or survival. The appearance of a new generation of potentially effective tools in the early 1980s, including hydroabsorbent polymers and copolymers from the propenamide and propenamide-propenoate families, opened new perspectives.

Application

Soil conditioners may be applied in a number of ways. Some are worked into the soil with a tiller before planting. Others are applied after planting, or periodically during the growing season. Soil testing should be performed prior to applying a soil conditioner to learn more about the composition and structure of the soil. This testing will determine which conditioners will be more appropriate for the conditions.

Ecological Concerns

While adding a soil conditioner to crops or a garden can seem like a great way to get healthier plants, over-application of some amendments can cause ecological problems. For example, salts, nitrogen, metals and other nutrients that are present in many soil amendments are not productive when added in excess, and can actually be detrimental to plant health. Runoff of excess nutrients into waterways also occurs, which is harmful to the water quality and through it, the environment.

Monoculture

Monoculture is the agricultural practice of producing or growing a single crop, plant, or livestock species, variety, or breed in a field or farming system at a time. Polyculture, where more than one crop is grown in the same space at the same time, is the alternative to monoculture. Monoculture is widely used in both industrial farming and organic farming and has allowed increased efficiency in planting and harvest.

Continuous monoculture, or monocropping, where the same species is grown year after year, can lead to the quicker buildup of pests and diseases, and then rapid spread where a uniform crop is susceptible to a pathogen. The practice has increasingly come under fire for its environmental effects and for putting the food supply chain at risk. Diversity can be added both in time, as with a crop rotation or sequence, or in space, with a polyculture.

Diversity of crops in space and time; monocultures and polycultures, and rotations of both.					
			Diversity in time		
				Higher	
			Low	Cyclic	Dynamic (non-cyclic)
Diversity in space	**Low**	Monoculture, one species in a field	Continuous monoculture, monocropping	Crop rotation (rotation of monocultures)	Sequence of monocultures
	Higher	Polyculture, two or more species intermingled in a field	Continuous polyculture	Rotation of polycultures	Sequence of polycultures

Oligoculture has been suggested to describe a crop rotation of just a few crops, as is practiced by several regions of the world.

The term monoculture is frequently applied for other uses to describe any group dominated by a single variety, e.g. social Monoculturalism, or in the field of musicology to describe the dominance of the American and British music-industries in Western pop music, or in the field of computer science to describe a group of computers all running identical software.

A monocultivated potato field

Land Use

The term is mostly used in agriculture and describes the practice of planting the same crop species and cultivar in a field. Examples include wheat fields or apple orchards or grape vineyards. Because each species has the same growing requirements and habits, planting, maintenance (including pest control), and harvesting can be standardized, resulting in greater yields. It also is beneficial because a crop can be tailor-planted for a location that has special problems – like soil salt or drought or a short growing season.

When a crop is matched to its well-managed environment, a monoculture can produce higher yields than a polyculture. In the last 40 years, modern practices such as monoculture planting and the use of synthesized fertilizers have reduced the amount of additional land needed to produce food. However, planting the same crop in the same place each year depletes the nutrients from the earth that the plant relies on and leaves soil weak and unable to support healthy plant growth. Because soil structure and quality is so poor, farmers are forced to use chemical fertilizers to encourage plant growth and fruit production. These fertilizers, in turn, disrupt the natural makeup of the soil and contribute further to nutrient depletion. Monocropping also creates the spread of pests and diseases, which have to be treated with yet more chemicals. The effects of monocropping on the environment are severe when pesticides and fertilizers make their way into ground water or become airborne, creating pollution.

Forestry

In forestry, monoculture refers to the planting of one species of tree. Monoculture plantings provide great yields and more efficient harvesting than natural stands of trees. Single-species stands of trees are often the natural way trees grow, but the stands show a diversity in tree sizes, with dead trees mixed with mature and young trees. In forestry, monoculture stands that are planted and harvested as a unit provide limited resources for wildlife that depend on dead trees and openings, since all the trees are the same size; they are most often harvested by clearcutting, which drastically alters the habitat. The mechanical harvesting of trees can compact soils, which can adversely affect understory growth. Single-species planting of trees also are more vulnerable when infected with a pathogen, or are attacked by insects, and by adverse environmental conditions.

Lawns and Animals

Examples of monoculture include lawns and most field cs wheat or corn. The term is also used where a single breed of farm animal is raised in large-scale concentrated animal feeding operations (CAFOs). In the United States, The Livestock Conservancy was formed to protect nearly 200 endangered livestock breeds from going extinct, largely due to the increased reliance on just a handful of highly specialized breeds.

Disease

Crops used in agriculture are usually single strains that have been bred for high yield and resistant to certain common diseases. Since all plants in a monoculture are genetically similar, if a disease strikes to which they have no resistance, it can destroy entire populations of crops. Polyculture, which is the mixing of different crops, reduces the likelihood that one or more of the crops will be resistant to any particular pathogen. Studies have shown planting a mixture of crop strains in the same field to be effective at combating disease. Ending monocultures grown under disease conditions by introducing crop diversity has greatly increased yields. In one study in China, the planting of several varieties of rice in the same field increased yields of non-resistant strains by 89% compared to non-resistant strains grown in monoculture, largely because of a dramatic (94%) decrease in the incidence of disease, making pesticides less necessary. There is currently a great deal of international worry about the wheat leaf rust fungus, that has already decimated wheat crops in Uganda and Kenya, and is starting to make inroads into Asia as well. As much of the world's wheat crops are very genetically similar following the Green Revolution, the impacts of such diseases threaten agricultural production worldwide.

In Ireland, exclusive use of one variety of potato, the "Lumper", lead to the great famine. It was cheap food to feed the masses. Potatoes were propagated vegetatively so for all intents and purposes, the potatoes were clones with little to no genetic variation. When the potato fungus called "the Blight" arrived from the Americas in 1845 to Ireland, the Lumper had no resistance to the disease leading to the nearly complete failure of the potato crop across Ireland. Had the farmers used multiple varieties of potato, the famine may not have occurred. Keep in mind that Andean natives were cultivating three thousand varieties before the Spaniards arrived. Today in the US we cultivate 250 varieties.

Many of today's livestock production systems rely on just a handful of highly specialized breeds. By focusing heavily on a single trait (output), other traits like fertility, disease resistance, vigor, and mothering instincts are sacrificed. In the early 1990s a few Holstein calves were observed to grow poorly and died in the first 6 months of life. They were all found to be homozygous for a mutation in the gene that caused Bovine Leukocyte Adhesion Deficiency. This mutation was found at a high frequency in Holstein populations world-wide. (15% among bulls in the US, 10% in Germany, and 16% in Japan.) By studying the pedigrees of affected and carrier animals the source of the mutation was tracked to a single bull that was widely used in the industry. Note that in 1990 there were approximately 4 million Holteins in the US making the affected population around 600,000 animals.

Polyculture

The environmental movement seeks to change popular culture by redefining the "perfect lawn" to be something other than a turf monoculture, and seeks agricultural policy that provides greater encouragement for more diverse cropping systems. Local food systems may also encourage growing multiple species and a wide variety of crops at the same time and same place. Heirloom gardening and raising heritage livestock breeds have come about largely as a reaction against monocultures in agriculture.

Polyculture

Polyculture is agriculture using multiple crops in the same space, providing crop diversity in imitation of the diversity of natural ecosystems, and avoiding large stands of single crops, or monoculture. It includes multi-cropping, intercropping, companion planting, beneficial weeds, and alley cropping. It is the raising at the same time and place of more than one species of plant or animal. Polyculture is one of the principles of permaculture.

Polyculture providing useful within-field diversity: companion planting of carrots and onions. The onion smell puts off carrot root fly, while the smell of carrots puts off onion fly.

Advantages

Polyculture, though it often requires more labor, has two main advantages over monoculture.

Polyculture reduces susceptibility to disease. For example, a study in China showed that planting several varieties of rice in the same field increased yields by 89%, largely because of a dramatic (94%) decrease in the incidence of disease, which made pesticides redundant.

Polyculture increases local biodiversity. This is one example of reconciliation ecology, or accommodating biodiversity within human landscapes. This may also form part of a biological pest control program.

Cover Crop

A cover crop is a crop planted primarily to manage soil erosion, soil fertility, soil quality, water, weeds, pests, diseases, biodiversity and wildlife in an *agroecosystem* (Lu *et al.* 2000), an ecological system managed and largely shaped by humans across a range of intensities to produce food, feed, or fiber. Currently, not many countries are known for using the cover crop method.

Cover crops are of interest in sustainable agriculture as many of them improve the sustainability of agroecosystem attributes and may also indirectly improve qualities of neighboring natural ecosystems. Farmers choose to grow and manage specific cover crop types based on their own needs and goals, influenced by the biological, environmental, social, cultural, and economic factors of the food system in which farmers operate (Snapp *et al.* 2005). The farming practice of cover crops has been recognized as climate-smart agriculture by the White House.

Soil Erosion

Although cover crops can perform multiple functions in an agroecosystem simultaneously, they are often grown for the sole purpose of preventing soil erosion. Soil erosion is a process that can irreparably reduce the productive capacity of an agroecosystem. Dense cover crop stands physically slow down the velocity of rainfall before it contacts the soil surface, preventing soil splashing and erosive surface runoff (Romkens *et al.* 1990). Additionally, vast cover crop root networks help anchor the soil in place and increase soil porosity, creating suitable habitat networks for soil macrofauna (Tomlin *et al.* 1995).

Soil Fertility Management

One of the primary uses of cover crops is to increase soil fertility. These types of cover crops are referred to as "green manure." They are used to manage a range of soil macronutrients and micronutrients. Of the various nutrients, the impact that cover crops have on nitrogen management has received the most attention from researchers and farmers, because nitrogen is often the most limiting nutrient in crop production.

Often, green manure crops are grown for a specific period, and then plowed under before reaching full maturity in order to improve soil fertility and quality. Also the stalks left block the soil from being eroded.

Green manure crops are commonly leguminous, meaning they are part of the Fabaceae (pea) family.This family is unique in that all of the species in it set pods, such as bean, lentil, lupins and alfalfa. Leguminous cover crops are typically high in nitrogen and can often provide the required quantity of nitrogen for crop production. In conventional farming, this nitrogen is typically applied in chemical fertilizer form. This quality of cover crops is called fertilizer replacement value (Thiessen-Martens *et al.* 2005).

Another quality unique to leguminous cover crops is that they form symbiotic relationships with the rhizobial bacteria that reside in legume root nodules. Lupins is nodulated by the soil microorganism *Bradyrhizobium* sp. (Lupinus). Bradyrhizobia are encountered as microsymbionts in other leguminous crops (*Argyrolobium, Lotus, Ornithopus, Acacia, Lupinus*) of Mediterranean origin. These bacteria convert biologically unavailable atmospheric nitrogen gas (N

2) to biologically available ammonium (NH+

4) through the process of biological nitrogen fixation.

Prior to the advent of the Haber-Bosch process, an energy-intensive method developed to carry out industrial nitrogen fixation and create chemical nitrogen fertilizer, most nitrogen introduced to ecosystems arose through biological nitrogen fixation (Galloway *et al.* 1995). Some scientists believe that widespread biological nitrogen fixation, achieved mainly through the use of cover crops, is the only alternative to industrial nitrogen fixation in the effort to maintain or increase future food production levels (Bohlool *et al.* 1992, Peoples and Craswell

1992, Giller and Cadisch 1995). Industrial nitrogen fixation has been criticized as an unsustainable source of nitrogen for food production due to its reliance on fossil fuel energy and the environmental impacts associated with chemical nitrogen fertilizer use in agriculture (Jensen and Hauggaard-Nielsen 2003). Such widespread environmental impacts include nitrogen fertilizer losses into waterways, which can lead to eutrophication (nutrient loading) and ensuing hypoxia (oxygen depletion) of large bodies of water.

An example of this lies in the Mississippi Valley Basin, where years of fertilizer nitrogen loading into the watershed from agricultural production have resulted in a hypoxic "dead zone" off the Gulf of Mexico the size of New Jersey (Rabalais *et al.* 2002). The ecological complexity of marine life in this zone has been diminishing as a consequence (CENR 2000).

As well as bringing nitrogen into agroecosystems through biological nitrogen fixation, types of cover crops known as "catch crops" are used to retain and recycle soil nitrogen already present. The catch crops take up surplus nitrogen remaining from fertilization of the previous crop, preventing it from being lost through leaching (Morgan *et al.* 1942), or gaseous denitrification or volatilization (Thorup-Kristensen *et al.* 2003).

Catch crops are typically fast-growing annual cereal species adapted to scavenge available nitrogen efficiently from the soil (Ditsch and Alley 1991). The nitrogen tied up in catch crop biomass is released back into the soil once the catch crop is incorporated as a green manure or otherwise begins to decompose.

An example of green manure use comes from Nigeria, where the cover crop *Mucuna pruriens* (velvet bean) has been found to increase the availability of phosphorus in soil after a farmer applies rock phosphate (Vanlauwe *et al.* 2000).

Soil Quality Management

Cover crops can also improve soil quality by increasing soil organic matter levels through the input of cover crop biomass over time. Increased soil organic matter enhances soil structure, as well as the water and nutrient holding and buffering capacity of soil (Patrick *et al.* 1957). It can also lead to increased soil carbon sequestration, which has been promoted as a strategy to help offset the rise in atmospheric carbon dioxide levels (Kuo *et al.* 1997, Sainju *et al.* 2002, Lal 2003).

Soil quality is managed to produce optimum circumstances for crops to flourish. The principal factors of soil quality are soil salination, pH, microorganism balance and the prevention of soil contamination.

Water Management

By reducing soil erosion, cover crops often also reduce both the rate and quantity of water that drains off the field, which would normally pose environmental risks to waterways and ecosystems downstream (Dabney *et al.* 2001). Cover crop biomass acts as a physical barrier between rainfall and the soil surface, allowing raindrops to steadily trickle down through the soil profile. Also, as stated above, cover crop root growth results in the formation of soil pores, which in addition to enhancing soil macrofauna habitat provides pathways for water to filter through the soil profile rather than draining off the field as surface flow. With increased water infiltration, the potential for soil water storage and the recharging of aquifers can be improved (Joyce *et al.* 2002).

Just before cover crops are killed (by such practices including mowing, tilling, discing, rolling, or herbicide application) they contain a large amount of moisture. When the cover crop is incorporated into the soil, or left on the soil surface, it often increases soil moisture. In agroecosystems where water for crop production is in short supply, cover crops can be used as a mulch to conserve water by shading and cooling the soil surface. This reduces evaporation of soil moisture. In other situations farmers try to dry the soil out as quickly as possible going into the planting season. Here prolonged soil moisture conservation can be problematic.

While cover crops can help to conserve water, in temperate regions (particularly in years with below average precipitation) they can draw down soil water supply in the spring, particularly if climatic growing conditions are good. In these cases, just before crop planting, farmers often face a tradeoff between the benefits of increased cover crop growth and the drawbacks of reduced soil moisture for cash crop production that season. C/N ratio is balanced with this application.

Weed Management

Cover crop in South Dakota

Thick cover crop stands often compete well with weeds during the cover crop growth period, and can prevent most germinated weed seeds from completing their life cycle and reproducing. If the cover crop is left on the soil surface rather than incorporated into the soil as a green manure after its growth is terminated, it can form a nearly impenetrable mat. This drastically reduces light transmittance to weed seeds, which in many cases reduces weed seed germination rates (Teasdale 1993). Furthermore, even when weed seeds germinate, they often run out of stored energy for growth before building the necessary structural capacity to break through the cover crop mulch layer. This is often termed the *cover crop smother effect* (Kobayashi *et al.* 2003).

Some cover crops suppress weeds both during growth and after death (Blackshaw *et al.* 2001). During growth these cover crops compete vigorously with weeds for available space, light, and nutrients, and after death they smother the next flush of weeds by forming a mulch layer on the soil surface. For example, Blackshaw *et al.* (2001) found that when using *Melilotus officinalis* (yellow sweetclover) as a cover crop in an improved fallow system (where a fallow period is intentionally improved by any number of different management practices, including the planting of cover crops), weed biomass only constituted between 1-12% of total standing biomass at the end of the cover crop growing season. Furthermore, after cover crop termination, the yellow sweetclover residues suppressed weeds to levels 75-97% lower than in fallow (no yellow sweetclover) systems .

In addition to competition-based or physical weed suppression, certain cover crops are known to suppress weeds through allelopathy (Creamer *et al.* 1996, Singh *et al.* 2003). This occurs when certain biochemical cover crop compounds are degraded that happen to be toxic to, or inhibit seed germination of, other plant species. Some well known examples of allelopathic cover crops are *Secale cereale* (rye), *Vicia villosa* (hairy vetch), *Trifolium pratense* (red clover), *Sorghum bicolor* (sorghum-sudangrass), and species in the Brassicaceae family, particularly mustards (Haramoto and Gallandt 2004). In one study, rye cover crop residues were found to have provided between 80% and 95% control of early season broadleaf weeds when used as a mulch during the production of different cash crops such as soybean, tobacco, corn, and sunflower (Nagabhushana *et al.* 2001).

In a recent study released by the Agricultural Research Service (ARS) scientists examined how rye seeding rates and planting patterns affected cover crop production. [1] The results show that planting more pounds per acre of rye increased the cover crop's production as well as decreased the amount of weeds. The same was true when scientists tested seeding rates on legumes and oats; a higher density of seeds planted per acre decreased the amount of weeds and increased the yield of legume and oat production. The planting patterns, which consisted of either traditional rows or grid patterns, did not seem to make a significant impact on the cover crop's production or on the weed production in either cover crop. The ARS scientists concluded that increased seeding rates could be an effective method of weed control.

Disease Management

In the same way that allelopathic properties of cover crops can suppress weeds, they can also break disease cycles and reduce populations of bacterial and fungal diseases (Everts 2002), and parasitic nematodes (Potter *et al.* 1998, Vargas-Ayala *et al.* 2000). Species in the Brassicaceae family, such as mustards, have been widely shown to suppress fungal disease populations through the release of naturally occurring toxic chemicals during the degradation of glucosinolade compounds in their plant cell tissues (Lazzeri and Manici 2001).

Pest Management

Some cover crops are used as so-called "trap crops", to attract pests away from the crop of value and toward what the pest sees as a more favorable habitat (Shelton and Badenes-Perez 2006). Trap crop areas can be established within crops, within farms, or within landscapes. In many cases the trap crop is grown during the same season as the food crop being produced. The limited area occupied by these trap crops can be treated with a pesticide once pests are drawn to the trap in large enough numbers to reduce the pest populations. In some organic systems, farmers drive over the trap crop with a large vacuum-based implement to physically pull the pests off the plants and out of the field (Kuepper and Thomas 2002). This system has been recommended for use to help control the lygus bugs in organic strawberry production (Zalom *et al.* 2001). Another example of trap crops are nematode resistance White mustard (*Sinapis alba*) and Radish (*Raphanus sativus*). They can be grown after a main (cereal) crop and trap nematodes, for example the beet cyst nematode and Columbian root knot nematode. When grown, nematodes hatch and are attracted to the roots. After entering the roots they cannot reproduce in the root due to a hypersensitive resistance reaction of the plant. Hence the nematode population is greatly reduced, by 70-99%, depending on species and cultivation time.

Other cover crops are used to attract natural predators of pests by providing elements of their habitat. This is a form of biological control known as habitat augmentation, but achieved with the use of cover crops (Bugg and Waddington 1994). Findings on the relationship between cover crop presence and predator/pest population dynamics have been mixed, pointing toward the need for detailed information on specific cover crop types and management practices to best complement a given integrated pest management strategy. For example, the predator mite *Euseius tularensis* (Congdon) is known to help control the pest citrus thrips in Central California citrus orchards. Researchers found that the planting of several different leguminous cover crops (such as bell bean, woollypod vetch, New Zealand white clover, and Austrian winter pea) provided sufficient pollen as a feeding source to cause a seasonal increase in *E. tularensis* populations, which with good timing could potentially introduce enough predatory pressure to reduce pest populations of citrus thrips (Grafton-Cardwell *et al.* 1999).

Diversity and Wildlife

Although cover crops are normally used to serve one of the above discussed purposes, they often simultaneously improve farm habitat for wildlife. The use of cover crops adds at least one more dimension of plant diversity to a cash crop rotation. Since the cover crop is typically not a crop of value, its management is usually less intensive, providing a window of "soft" human influence on the farm. This relatively "hands-off" management, combined with the increased on-farm heterogeneity created by the establishment of cover crops, increases the likelihood that a more complex trophic structure will develop to support a higher level of wildlife diversity (Freemark and Kirk 2001).

In one study, researchers compared arthropod and songbird species composition and field use between conventionally and cover cropped cotton fields in the Southern United States. The cover cropped cotton fields were planted to clover, which was left to grow in between cotton rows throughout the early cotton growing season (stripcover cropping). During the migration and breeding season, they found that songbird densities were 7–20 times higher in the cotton fields with integrated clover cover crop than in the conventional cotton fields. Arthropod abundance and biomass was also higher in the clover cover cropped fields throughout much of the songbird breeding season, which was attributed to an increased supply of flower nectar from the clover. The clover cover crop enhanced songbird habitat by providing cover and nesting sites, and an increased food source from higher arthropod populations (Cederbaum *et al.* 2004).

Soil Steam Sterilization

Soil steam sterilization (soil steaming) is a farming technique that sterilizes soil with steam in open fields or greenhouses. Pests of plant cultures such as weeds, bacteria, fungi and viruses are killed through induced hot steam which causes their cell structure to physically degenerate. Biologically, the method is considered a partial disinfection. Important heat-resistant, spore-forming bacteria survive and revitalize the soil after cooling down. Soil fatigue can be cured through the release of nutritive substances blocked within the soil. Steaming leads to a better starting position, quicker growth and strengthened resistance against plant disease and pests. Today, the application of hot steam is considered the best and most effective way to disinfect sick soil, potting soil and compost. It is being used as an alternative to bromomethane, whose production and use was curtailed by the Montreal Protocol. "Steam effectively kills pathogens by heating the soil to levels that cause protein coagulation or enzyme inactivation."

Benefits of Soil Steaming

Soil sterilization provides secure and quick relief of soils from substances and organisms harmful to plants such as:

- Bacteria

- Viruses

- Fungi

- Nematodes and

- Other Pests

Further positive effects are:

- All weed and weed seeds are killed

- Significant increase of crop yields

- Relief from soil fatigue through activation of chemical – biological reactions

- Blocked nutritive substances in the soil are tapped and made available for plants

- Alternative to Methyl Bromide and other critical chemicals in agriculture

Steaming with Superheated Steam

Through modern steaming methods with superheated steam at 180–200 °C, an optimal soil disinfection can be achieved. Soil only absorbs a small amount of humidity. Micro organisms become active once the soil has cooled down. This creates an optimal environment for instant tillage with seedlings and seeds. Additionally the method of integrated steaming can promote a target-oriented resettlement of steamed soil with beneficial organisms. In the process, the soil is first freed from all organisms and then revitalized and microbiologically buffered through the injection of a soil activator based on compost which contains a natural mixture of favorable microorganisms (e.g. Bacillus subtilis, etc.).

Different types of such steam application are also available in practice, including substrate steaming and surface steaming.

Surface Steaming

Several methods for surface steaming are in use amongst which are: area sheet steaming, the steaming hood, the steaming harrow, the steaming plough and vacuum steaming with drainage pipes or mobile pipe systems.

In order to pick the most suitable steaming method, certain factors have to be considered such as soil structure, plant culture and area performance. At present, more advanced methods are being developed, such as sandwich steaming or partially integrated sandwich steaming in order to minimize energy and cost as much as possible.

Sheet Steaming

Sheet steaming with a MSD/moeschle steam boiler (left side)

Large area sheet steaming in greenhouses using a steam injector.

Surface steaming with special sheets (sheet steaming) is a method which has been established for decades in order to steam large areas reaching from 15 to 400 m² in one step. If properly applied, sheet steaming is simple and highly economic. The usage of heat resistant, non-decomposing insulation fleece saves up to 50% energy, reduces the steaming time significantly and improves penetration. Single working step areas up to 400 m² can be steamed in 4–5 hours down to 25–30 cm depth / 90°C. The usage of heat resistant and non-decomposing synthetic insulation fleece, 5 mm thick, 500 gr / m², can reduce steaming time by about 30%. Through a steam injector or a perforated pipe, steam is injected underneath the sheet after it has been laid out and weighted with sand sacks.

The area performance in one working step depends on the capacity of the steam generator (e.g. steam boiler):

Steam capacity kg/h:	100	250	300	400	550	800	1000	1350	2000
Area m²:	15-20	30-50	50-65	60-90	80-120	130-180	180-220	220-270	300-400

The steaming time depends on soil structure as well as outside temperature and amounts to 1-1.5 hours per 10 cm steaming depth. Hereby the soil reaches a temperature of about 85°C. Milling for soil loosening is not recommended since soil structure may become too fine which reduces its penetrability for steam. The usage of spading machines is ideal for soil loosening. The best results can be achieved if the soil is cloddy at greater depth and granulated at lesser depth.

In practice, working with at least two sheets simultaneously has proven to be highly effective. While one sheet is used for steaming the other one is prepared for steam injection, therefore unnecessary steaming recesses are avoided.

Depth Steaming with Vacuum

Steaming with vacuum which is induced through a mobile or fixed installed pipe system in the depth of the area to be steamed, is the method that reaches the best penetration. Despite high capital cost, the fixed installation of drainage systems is reasonable for intensively used areas since steaming depths of up to 80 cm can be achieved.

In contrast to fixed installed drainage systems, pipes in mobile suction systems are on the surface. A central suction pipeline consisting of zinc-coated, fast-coupling pipes are connected in a regular spacing of 1.50 m and the ends of the hoses are pushed into the soil to the desired depth with a special tool.

The steaming area is covered with a special steaming sheet and weighted all around as with sheet steaming. The steam is injected underneath the sheet through an injector and protection tunnel. While with short areas up to 30 m length steam is frontally injected, with longer areas steam is induced in the middle of the beet using a T-connection branching out to both sides. As soon as the sheet is inflated to approximately 1m by the steam pressure, the suction turbine is switched on. First, the air in the soil is removed via the suction hoses. A vacuum is formed and the steam is pulled downward.

During the final phase, when the required steaming depth has been reached, the ventilator runs non-stop and surplus steam is blown out. To ensure that this surplus steam is not lost, it is fed back under the sheet.

As with all other steaming systems, a post-steaming period of approximately 20–30 minutes is required. Steaming time is approximately 1 hour per 10 cm steaming depth. The steam requirement is approximately 7–8 kg/m².

The most important requirement, as with all steaming systems, is that the soil is well loosened before steaming, to ensure optimal penetration.

Negative Pressure Technique

"Negative Pressure technique generates appropriate soil temperature at a 60 cm depth and complete control of nematodes, fungi and weeds is achieved. In this technique, the steam is introduced under the steaming sheath and forced to enter the soil profile by a negative pressure. The negative pressure is created by a fan that sucks the air out of the soil through buried perforated polypropylene pipes. This system requires a permanent installation of perforated pipes into the soil, at a depth of at least 60 cm to be protected from plough."

Steaming with Hoods

Half automatic steaming hood with three wings in greenhouse

A steaming hood is a mobile device consisting of corrosion-resistant materials such as aluminum, which is put down onto the area to be steamed. In contrast to sheet steaming, cost-intensive working steps such as laying out and weighting the sheets don't occur, however the area steamed per working step is smaller in accordance to the size of the hood.

Outdoors, a hood is positioned either manually or via tractor with a special pre-stressed 4 point suspension arm. Steaming time amounts to 30 min for a penetration down to 25 cm depth. Hereby a temperature of 90°C can be reached. In large stable glasshouses, the hoods are attached to tracks. They are lifted and moved by pneumatic cylinders. Small and medium-sized hoods up to 12m² are lifted manually using a tipping lever or moved electrically with special winches.

Combined Surface and Depth Injection of Steam (Sandwich Steaming)

Sandwich steaming, which was developed in a project among DEIAFA, University of Turin (Italy, www.deiafa.unito.it) and Ferrari Costruzioni Meccaniche, represents a combination of depth and surface steaming, offers an efficient method to induce hot steam into the soil. The steam is simultaneously pushed into the soil from the surface and from the depth. For this purpose, the area, which must be equipped with a deep steaming injection system, is covered with a steaming hood. The steam enters the soil from the top and the bottom at the same time. Sheets are not suitable, since a high pressure up to 30 mm water column arises underneath the cover.

Sandwich steaming machine model Sterilter constructed by Ferrari Costruzioni Meccaniche equipped with MSD/ moeschle steam boiler

Sandwich steaming offers several advantages. On the one hand, application of energy can be increased to up to 120 kg steam per m²/h. In comparison to other steaming methods up to 30% energy savings can be achieved and the usage of fuel (e.g. heating oil) accordingly decreases. The increased application of energy leads to a quick heating of the soil which reduces the loss of heat. On the other hand, only half of the regular steaming time is needed.

Comparison of sandwich steaming with other steam injection methods relating to steam output and energy demand(*):

Steaming method	max. steam output	energy demand (*)
Sheet steaming	6 kg/m²h	about 100 kg steam/m³
Depth steaming (Sheet + vacuum)	14 kg/m²h	about 120 kg steam/m³
Hood steaming (Alu)	30 kg/m²h	about 80 kg steam/m³
Hood steaming (Steel)	50 kg/m²h	about 75 kg steam/m³
Sandwich steaming	120 kg/m²h	about 60 kg steam/m³

(*) in soil max 30% moisture

Clearly, Sandwich steaming reaches the highest steam output at the lowest energy demand.

Partially Integrated Sandwich Steaming

The partial integrated Sandwich steaming is an advanced combined method for steaming merely the areas which shall be planted and purposely leaving out those areas which shall not be used. In order to avoid risk of re-infection of steamed areas with pest from unsteamed areas, beneficial organisms can directly be injected into the hygenized soil via a soil activator (e.g. special compost). The partial sandwich steaming unlocks further potential savings in the steaming process.

Container / Stack Steaming

Stack steaming is used when thermically treating compost and substrates such as turf. Depending on the amount, the material to be steamed is piled up to 70 cm height in steaming boxes or in small dump trailers. Steam is evenly injected via manifolds. For huge amounts, steaming containers and soil boxes are used which are equipped with suction systems to improve steaming results. Midget amounts can be steamed in special small steaming devices.

The amount of soil steamed should be tuned in a way that steaming time amounts to at most 1.5 h in order to avoid large quantities of condensed water in the bottom layers of the soil.

Steam Output kg/h:	100	250	300	400	550	800	1000	1350	2000
m³/h about:	1.0-1.5	2.5-3.0	3.0-3.5	4.0-5.0	5.5-7.0	8.0-10.0	10.0-13.0	14.0-18.0	20.0-25.0

In light substrates, such as turf, the performance per hour is significantly higher.

History

Ancient civilizations in India and Egypt used steam, generated through the targeted usage of incident solar radiation on watered top soil, to sanitize and revive their arable land.

Modern soil steam sterilization was first discovered in 1888 (by Frank in Germany) and was first commercially used in the United States (by Rudd) in 1893 (Baker 1962). Since then, a wide variety of steam machines have been built to disinfest both commercial greenhouse and nursery field soils (Grossman and Liebman 1995). In the 1950s, for example, steam sterilization technologies expanded from disinfestation of potting soil and greenhouse mixes to commercial production of steam rakes and tractor-drawn steam blades for fumigating small acres of cut flowers and other high-value field crops (Langedijk 1959). Today, even more effective steam technologies are being developed.

Application of Hot Steam

- In horticulture as well as nurseries for sterilization of substrates and top soil

- In agriculture for sterilization and treatment of food waste for pig fattening and heating of molasses

- In mushroom cultivation for pasteurization of growing rooms, sterilization of top soil and combined application as heating

- In wineries as combination boiler for sterilization and cleaning of storage tanks, tempering of mash and for warm water generation.

Permaculture

Permaculture is a system of agricultural and social design principles centered on simulating or directly utilizing the patterns and features observed in natural ecosystems. The term *permaculture* (as a systematic method) was first coined by David Holmgren, then a graduate student, and his professor, Bill Mollison, in 1978. The word *permaculture* originally referred to "permanent agriculture", but was expanded to stand also for "permanent culture", as it was understood that social aspects were integral to a truly sustainable system as inspired by Masanobu Fukuoka's natural farming philosophy.

It has many branches that include but are not limited to ecological design, ecological engineering, environmental design, construction and integrated water resources management that develops sustainable architecture, regenerative and self-maintained habitat and agricultural systems modeled from natural ecosystems.

Mollison has said: "Permaculture is a philosophy of working with, rather than against nature; of protracted and thoughtful observation rather than protracted and thoughtless labor; and of looking at plants and animals in all their functions, rather than treating any area as a single product system."

History

In 1929, Joseph Russell Smith took up an antecedent term as the subtitle for *Tree Crops: A Permanent Agriculture*, a book in which he summed up his long experience experimenting with fruits and nuts as crops for human food and animal feed. Smith saw the world as an inter-related whole and suggested mixed systems of trees and crops underneath. This book inspired many individuals intent on making agriculture more sustainable, such as Toyohiko Kagawa who pioneered forest farming in Japan in the 1930s.

The definition of permanent agriculture as that which can be sustained indefinitely was supported by Australian P. A. Yeomans in his 1964 book *Water for Every Farm*. Yeomans introduced an observation-based approach to land use in Australia in the 1940s, and the keyline design as a way of managing the supply and distribution of water in the 1950s.

Stewart Brand's works were an early influence noted by Holmgren. Other early influences include Ruth Stout and Esther Deans, who pioneered no-dig gardening, and Masanobu Fukuoka who, in the late 1930s in Japan, began advocating no-till orchards, gardens and natural farming.

Core Tenets and Principles of Design

The three core tenets of permaculture are:

- ***Care for the earth***: Provision for all life systems to continue and multiply. This is the first principle, because without a healthy earth, humans cannot flourish.

- ***Care for the people***: Provision for people to access those resources necessary for their existence.

- ***Return of surplus***: Reinvesting surpluses back into the system to provide for the first two ethics. This includes returning waste back into the system to recycle into usefulness. The third ethic is sometimes referred to as Fair Share to reflect that each of us should take no more than what we need before we reinvest the surplus.

Permaculture design emphasizes patterns of landscape, function, and species assemblies. It determines where these elements should be placed so they can provide maximum benefit to the local environment. The central concept of permaculture is maximizing useful connections between components and synergy of the final design. The focus of permaculture, therefore, is not on each separate element, but rather on the relationships created among elements by the way they are placed together; the whole becoming greater than the sum of its parts. Permaculture design therefore seeks to minimize waste, human labor, and energy input by building systems with maximal benefits between design elements to achieve a high level of synergy. Permaculture designs evolve over time by taking into account these relationships and elements and can become extremely complex systems that produce a high density of food and materials with minimal input.

The design principles which are the conceptual foundation of permaculture were derived from the science of systems ecology and study of pre-industrial examples of sustainable land use. Permaculture draws from several disciplines including organic farming, agroforestry, integrated farming, sustainable development, and applied ecology. Permaculture has been applied most commonly to the design of housing and landscaping, integrating techniques such as agroforestry, natural building, and rainwater harvesting within the context of permaculture design principles and theory.

Theory

Twelve Design Principles

Twelve Permaculture design principles articulated by David Holmgren in his *Permaculture: Principles and Pathways Beyond Sustainability*:

1. *Observe and interact*: By taking time to engage with nature we can design solutions that suit our particular situation.

2. *Catch and store energy*: By developing systems that collect resources at peak abundance, we can use them in times of need.

3. *Obtain a yield*: Ensure that you are getting truly useful rewards as part of the work that you are doing.

4. *Apply self-regulation and accept feedback*: We need to discourage inappropriate activity to ensure that systems can continue to function well.

5. *Use and value renewable resources and services*: Make the best use of nature's abundance to reduce our consumptive behavior and dependence on non-renewable resources.

6. *Produce no waste*: By valuing and making use of all the resources that are available to us, nothing goes to waste.

7. *Design from patterns to details*: By stepping back, we can observe patterns in nature and society. These can form the backbone of our designs, with the details filled in as we go.

8. *Integrate rather than segregate*: By putting the right things in the right place, relationships develop between those things and they work together to support each other.

9. *Use small and slow solutions*: Small and slow systems are easier to maintain than big ones, making better use of local resources and producing more sustainable outcomes.

10. *Use and value diversity*: Diversity reduces vulnerability to a variety of threats and takes advantage of the unique nature of the environment in which it resides.

11. *Use edges and value the marginal*: The interface between things is where the most interesting events take place. These are often the most valuable, diverse and productive elements in the system.

12. *Creatively use and respond to change*: We can have a positive impact on inevitable change by carefully observing, and then intervening at the right time.

Layers

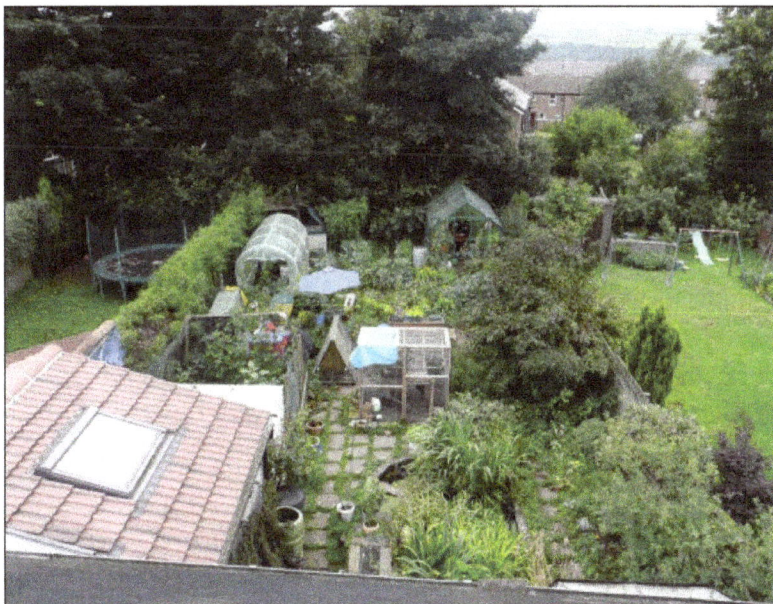

Suburban permaculture garden in Sheffield, UK with different layers of vegetation

Layers are one of the tools used to design functional ecosystems that are both sustainable and of direct benefit to humans. A mature ecosystem has a huge number of relationships between its component parts: trees, understory, ground cover, soil, fungi, insects, and animals. Because plants grow to different heights, a diverse community of life is able to grow in a relatively small space, as the vegetation occupies different layers. There are generally seven recognized layers in a food forest, although some practitioners also include fungi as an eighth layer.

1. The canopy: the tallest trees in the system. Large trees dominate but typically do not saturate the area, i.e. there exist patches barren of trees.

2. Understory layer: trees that revel in the dappled light under the canopy.

3. Shrub layer: a diverse layer of woody perennials of limited height. includes most berry bushes.

4. Herbaceous layer: Plants in this layer die back to the ground every winter (if winters are cold enough, that is). They do not produce woody stems as the Shrub layer does. Many culinary and medicinal herbs are in this layer. A large variety of beneficial plants fall into this layer. May be annuals, biennials or perennials

5. Soil surface/Groundcover: There is some overlap with the Herbaceous layer and the Groundcover layer; however plants in this layer grow much closer to the ground, grow densely to fill bare patches of soil, and often can tolerate some foot traffic. Cover crops retain soil and lessen erosion, along with green manures that add nutrients and organic matter to the soil, especially nitrogen

6. Rhizosphere: Root layers within the soil. The major components of this layer are the soil and the organisms that live within it such as plant roots (including root crops such as potatoes and other edible tubers), fungi, insects, nematodes, worms, etc.

7. Vertical layer: climbers or vines, such as runner beans and lima beans (vine varieties)

Guilds

There are many forms of guilds, including guilds of plants with similar functions (that could interchange within an ecosystem), but the most common perception is that of a mutual support guild. Such a guild is a group of species where each provides a unique set of diverse functions that work in conjunction, or harmony. Mutual support guilds are groups of plants, animals, insects, etc. that work well together. Some plants may be grown for food production, some have tap roots that draw nutrients up from deep in the soil, some are nitrogen-fixing legumes, some attract beneficial insects, and others repel harmful insects. When grouped together in a mutually beneficial arrangement, these plants form a guild.

Edge Effect

The edge effect in ecology is the effect of the juxtaposition or placing side by side of contrasting environments on an ecosystem. Permaculturists argue that, where vastly differing systems meet, there is an intense area of productivity and useful connections. An example of this is the coast; where the land and the sea meet there is a particularly rich area that meets a disproportionate percentage of human and animal needs. So this idea is played out in permacultural designs by using spirals in the herb garden or creating ponds that have wavy undulating shorelines rather than a simple circle or oval (thereby increasing the amount of edge for a given area).

Zones

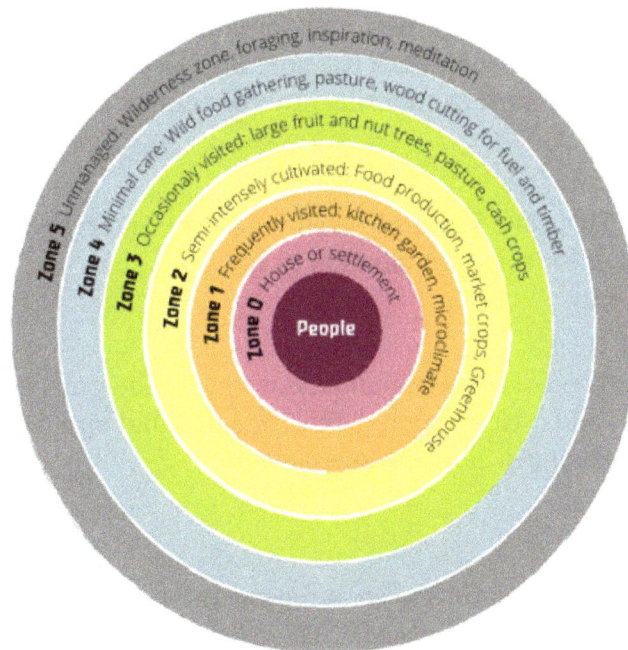

Permaculture Zones 0-5.

Zones are a way of intelligently organizing design elements in a human environment on the basis of the frequency of human use and plant or animal needs. Frequently manipulated or harvested elements of the design are located close to the house in zones 1 and 2. Less frequently used or manipulated elements, and elements that benefit from isolation (such as wild species) are farther away. Zones are about positioning things appropriately, and are numbered from 0 to 5.

Zone 0

> The house, or home center. Here permaculture principles would be applied in terms of aiming to reduce energy and water needs, harnessing natural resources such as sunlight, and generally creating a harmonious, sustainable environment in which to live and work. Zone 0 is an informal designation, which is not specifically defined in Bill Mollison's book.

Zone 1

> The zone nearest to the house, the location for those elements in the system that require frequent attention, or that need to be visited often, such as salad crops, herb plants, soft fruit like strawberries or raspberries, greenhouse and cold frames, propagation area, worm compost bin for kitchen waste, etc. Raised beds are often used in zone 1 in urban areas.

Zone 2

> This area is used for siting perennial plants that require less frequent maintenance, such

as occasional weed control or pruning, including currant bushes and orchards, pumpkins, sweet potato, etc. This would also be a good place for beehives, larger scale composting bins, and so on.

Zone 3

The area where main-crops are grown, both for domestic use and for trade purposes. After establishment, care and maintenance required are fairly minimal (provided mulches and similar things are used), such as watering or weed control maybe once a week.

Zone 4

A semi-wild area. This zone is mainly used for forage and collecting wild food as well as production of timber for construction or firewood.

Zone 5

A wilderness area. There is no human intervention in zone 5 apart from the observation of natural ecosystems and cycles. Through this zone we build up a natural reserve of bacteria, moulds and insects that can aid the zones above it.

People and Permaculture

Permaculture uses observation of nature to create regenerative systems, and the place where this has been most visible has been on the landscape. There has been a growing awareness though that firstly, there is the need to pay more attention to the peoplecare ethic, as it is often the dynamics of people that can interfere with projects, and secondly that the principles of permaculture can be used as effectively to create vibrant, healthy and productive people and communities as they have been in landscapes.

Domesticated Animals

Domesticated animals are often incorporated into site design.

Common Practices

Agroforestry

Agroforestry is an integrated approach of using the interactive benefits from combining trees and shrubs with crops and/or livestock. It combines agricultural and forestry technologies to create more diverse, productive, profitable, healthy and sustainable land-use systems. In agroforestry systems, trees or shrubs are intentionally used within agricultural systems, or non-timber forest products are cultured in forest settings.

Forest gardening is a term permaculturalists use to describe systems designed to mimic natural forests. Forest gardens, like other permaculture designs, incorporate processes and relationships that the designers understand to be valuable in natural ecosystems. The terms forest garden and food forest are used interchangeably in the permaculture literature. Numerous permaculturists are proponents of forest gardens, such as Graham Bell, Patrick Whitefield,

Dave Jacke, Eric Toensmeier and Geoff Lawton. Bell started building his forest garden in 1991 and wrote the book *The Permaculture Garden* in 1995, Whitefield wrote the book *How to Make a Forest Garden* in 2002, Jacke and Toensmeier co-authored the two volume book set *Edible Forest Gardening* in 2005, and Lawton presented the film *Establishing a Food Forest* in 2008.

Tree Gardens, such as Kandyan tree gardens, in South and Southeast Asia, are often hundreds of years old. Whether they derived initially from experiences of cultivation and forestry, as is the case in agroforestry, or whether they derived from an understanding of forest ecosystems, as is the case for permaculture systems, is not self-evident. Many studies of these systems, especially those that predate the term permaculture, consider these systems to be forms of agroforestry. Permaculturalists who include existing and ancient systems of polycropping with woody species as examples of food forests may obscure the distinction between permaculture and agroforestry.

Food forests and agroforestry are parallel approaches that sometimes lead to similar designs.

Hügelkultur

Hügelkultur is the practice of burying large volumes of wood to increase soil water retention. The porous structure of wood acts as a sponge when decomposing underground. During the rainy season, masses of buried wood can absorb enough water to sustain crops through the dry season. This technique has been used by permaculturalists Sepp Holzer, Toby Hemenway, Paul Wheaton and Masanobu Fukuoka.

Natural Building

A natural building involves a range of building systems and materials that place major emphasis on sustainability. Ways of achieving sustainability through natural building focus on durability and the use of minimally processed, plentiful or renewable resources, as well as those that, while recycled or salvaged, produce healthy living environments and maintain indoor air quality.

The basis of natural building is the need to lessen the environmental impact of buildings and other supporting systems, without sacrificing comfort, health or aesthetics. To be more sustainable, natural building uses primarily abundantly available, renewable, reused or recycled materials. In addition to relying on natural building materials, the emphasis on the architectural design is heightened. The orientation of a building, the utilization of local climate and site conditions, the emphasis on natural ventilation through design, fundamentally lessen operational costs and positively impact the environment. Building compactly and minimizing the ecological footprint is common, as are on-site handling of energy acquisition, on-site water capture, alternate sewage treatment and water reuse.

Rainwater Harvesting

Rainwater harvesting is the accumulating and storing of rainwater for reuse before it reaches the aquifer. It has been used to provide drinking water, water for livestock, water for irrigation, as well as other typical uses. Rainwater collected from the roofs of houses and local institutions can make an important contribution to the availability of drinking water. It can supplement the subsoil water level and increase urban greenery. Water collected from the ground, sometimes from areas which are especially prepared for this purpose, is called stormwater harvesting.

Greywater is wastewater generated from domestic activities such as laundry, dishwashing, and bathing, which can be recycled on-site for uses such as landscape irrigation and constructed wetlands. Greywater is largely sterile, but not potable (drinkable). Greywater differs from water from the toilets which is designated sewage or blackwater, to indicate it contains human waste. Blackwater is septic or otherwise toxic and cannot easily be reused. There are, however, continuing efforts to make use of blackwater or human waste. The most notable is for composting through a process known as humanure; a combination of the words human and manure. Additionally, the methane in humanure can be collected and used similar to natural gas as a fuel, such as for heating or cooking, and is commonly referred to as biogas. Biogas can be harvested from the human waste and the remainder still used as humanure. Some of the simplest forms of humanure use include a composting toilet or an outhouse or dry bog surrounded by trees that are heavy feeders which can be coppiced for wood fuel. This process eliminates the use of a standard toilet with plumbing.

Sheet Mulching

In agriculture and gardening, mulch is a protective cover placed over the soil. Any material or combination can be used as mulch, such as stones, leaves, cardboard, wood chips, gravel, etc., though in permaculture mulches of organic material are the most common because they perform more functions. These include: absorbing rainfall, reducing evaporation, providing nutrients, increasing organic matter in the soil, feeding and creating habitat for soil organisms, suppressing weed growth and seed germination, moderating diurnal temperature swings, protecting against frost, and reducing erosion. Sheet mulching is an agricultural no-dig gardening technique that attempts to mimic natural processes occurring within forests. Sheet mulching mimics the leaf cover that is found on forest floors. When deployed properly and in combination with other Permacultural principles, it can generate healthy, productive and low maintenance ecosystems.

Sheet mulch serves as a "nutrient bank," storing the nutrients contained in organic matter and slowly making these nutrients available to plants as the organic matter slowly and naturally breaks down. It also improves the soil by attracting and feeding earthworms, slaters and many other soil micro-organisms, as well as adding humus. Earthworms "till" the soil, and their worm castings are among the best fertilizers and soil conditioners. Sheet mulching can be used to reduce or eliminate undesirable plants by starving them of light, and can be more advantageous than using herbicide or other methods of control.

Intensive Rotational Grazing

Grazing has long been blamed for much of the destruction we see in the environment. However, it has been shown that when grazing is modeled after nature, the opposite effect can be seen. Also known as cell grazing, managed intensive rotational grazing (MIRG) is a system of grazing in which ruminant and non-ruminant herds and/or flocks are regularly and systematically moved to fresh pasture, range, or forest with the intent to maximize the quality and quantity of forage growth. This disturbance is then followed by a period of rest which allows new growth. MIRG can be used with cattle, sheep, goats, pigs, chickens, rabbits, geese, turkeys, ducks and other animals depending on the natural ecological community that is being mimicked. Sepp Holzer and Joel Salatin have shown how the disturbance caused by the animals can be the spark needed to start ecological succession or prepare ground for planting. Allan Savory's holistic management technique has been likened to "a permaculture approach to rangeland management". One variation on MIRG that is gaining rapid popularity is called eco-grazing. Often used to either control invasives or re-establish native species, in eco-grazing the primary purpose of the animals is to benefit the environment and the animals can be, but are not necessarily, used for meat, milk or fiber.

Keyline Design

Keyline design is a technique for maximizing beneficial use of water resources of a piece of land developed in Australia by farmer and engineer P. A. Yeomans. The *Keyline* refers to a specific topographic feature linked to water flow which is used in designing the drainage system of the site.

Fruit Tree Management

> The no-pruning option is usually ignored by fruit experts, though often practised by default in people's back gardens! But it has its advantages. Obviously it reduces work, and more surprisingly it can lead to higher overall yields.
>
> — *Whitefield, Patrick, How to make a forest garden, p. 16*

Masanobu Fukuoka, as part of early experiments on his family farm in Japan, experimented with no-pruning methods, noting that he ended up killing many fruit trees by simply letting them go, which made them become convoluted and tangled, and thus unhealthy. Then he realised this is the difference between natural-form fruit trees and the process of change of tree form that results from abandoning previously-pruned unnatural fruit trees. He concluded that the trees should be raised all their lives without pruning, so they form healthy and efficient branch patterns that follow their natural inclination. This is part of his implementation of the Tao-philosophy of Wú wéi translated in part as no-action (against nature), and he described it as no unnecessary pruning, nature farming or "do-nothing" farming, of fruit trees, distinct from non-intervention or literal no-pruning. He ultimately achieved yields comparable to or exceeding standard/intensive practices of using pruning and chemical fertilisation.

Another proponent of the no, or limited, pruning method is Sepp Holzer who used the method in connection with Hügelkultur berms. He has successfully grown several varieties of fruiting trees at altitudes (approximately 9,000 feet (2,700 m)) far above their normal altitude, temperature, and snow load ranges. He notes that the Hügelkultur berms kept and/or generated enough heat to allow the roots to survive during alpine winter conditions. The point of having unpruned branches, he notes, was that the longer (more naturally formed) branches bend over under the snow load until they touched the ground, thus forming a natural arch against snow loads that would break a shorter, pruned, branch.

Compost Management

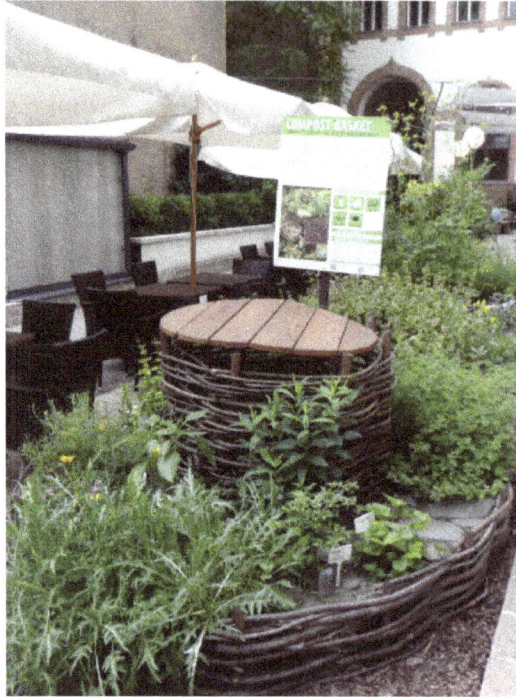

Compost Basket

Compost Basket is a way to the permanent management of compost materials. The idea of Compost Basket comes from Gyulai Iván. He invented and used it in the Gömörszőlős Educational Center. The inner circle is 1 meter deep. This is the place where compost material are put in. Under this is the outer circle, 40 cm deep. Here the nutrients come down.

Mollison and Holmgren

Bill Mollison in January 2008.

In the mid-1970s, Bill Mollison and David Holmgren started developing ideas about stable agricultural systems on the southern Australian island state of Tasmania. This was a result of the danger of the rapidly growing use of industrial-agricultural methods. In their view, highly dependent on non renewable resources, these methods were additionally poisoning land and water, reducing biodiversity, and removing billions of tons of topsoil from previously fertile landscapes. A design approach called *permaculture* was their response and was first made public with the publication of their book *Permaculture One* in 1978.

By the early 1980s, the concept had broadened from agricultural systems design towards sustainable human habitats. After *Permaculture One*, Mollison further refined and developed the ideas by designing hundreds of permaculture sites and writing more detailed books, notably *Permaculture: A Designers Manual*. Mollison lectured in over 80 countries and taught his two-week Permaculture Design Course (PDC) to many hundreds of students.Mollison "encouraged graduates to become teachers themselves and set up their own institutes and demonstration sites. This multiplier effect was critical to permaculture's rapid expansion."

In 1991, a four-part television documentary by ABC productions called "The Global Gardener" showed permaculture applied to a range of worldwide situations, bringing the concept to a much broader public. In 2012, the UMass Permaculture Initiative won the White House "Champions of Change" sustainability contest, which declared that "they demonstrate how permaculture can feed a growing population in an environmentally sustainable and socially responsible manner".

In 1997, Holmgren explained that the primary agenda of the permaculture movement is to assist people to become more self-reliant through the design and development of productive and sustainable gardens and farms.

In 2014, Holmgren endorsed and helped launch a new Australian permaculture magazine, Pip Magazine.

Notable Permaculturists

Joseph Russell Smith took up an antecedent term as the subtitle for *Tree Crops: A Permanent Agriculture*, a book in which he summed up his long experience experimenting with fruits and nuts as crops for human food and animal feed. By that year (1929), Smith saw the world as an interrelated whole and suggested mixed systems of trees and crops underneath. This book inspired many individuals intent on making permaculture a valid means of sustainable food production. Bill Mollison and David Holmgren developed it further, and permaculturists were trained under the umbrella of Bill Mollison's train the trainer system.

Geoff Lawton, Toby Hemenway and P. A. Yeomans - creator of the keyline design each have more than 20 years experience teaching and promoting permaculture as a sustainable way of growing food. Simon Fjell was a Founding Director of the Permaculture Institute in late 1979, over 40 years experience, having first met Mollison in 1976. He has since worked in every continent.

The permaculture movement also spread throughout Asia and Central America, with Hong Kong-based Asian Institute of Sustainable Architecture (AISA), Rony Lec leading the foundation of the Mesoamerican Permaculture Institute (IMAP) in Guatemala and Juan Rojas co-founding the Permaculture Institute of El Salvador.

Trademark and Copyright Issues

There has been contention over who, if anyone, controls legal rights to the word *permaculture*: is it trademarked or copyrighted? and if so, who holds the legal rights to the use of the word? For a long time Bill Mollison claimed to have copyrighted the word, and his books said on the copyright page, "The contents of this book and the word PERMACULTURE are copyright." These statements were largely accepted at face-value within the permaculture community. However, copyright law does not protect names, ideas, concepts, systems, or methods of doing something; it only protects the expression or the description of an idea, not the idea itself. Eventually Mollison acknowledged that he was mistaken and that no copyright protection existed for the word *permaculture*.

In 2000, Mollison's US based Permaculture Institute sought a service mark (a form of trademark) for the word *permaculture* when used in educational services such as conducting classes, seminars, or workshops. The service mark would have allowed Mollison and his two Permaculture Institutes (one in the US and one in Australia) to set enforceable guidelines regarding how permaculture could be taught and who could teach it, particularly with relation to the PDC, despite the fact that he had instituted a system of certification of teachers to teach the PDC in 1993. The service mark failed and was abandoned in 2001. Also in 2001 Mollison applied for trademarks in Australia for the terms "Permaculture Design Course" and "Permaculture Design". These applications were both withdrawn in 2003. In 2009 he sought a trademark for "Permaculture: A Designers' Manual" and "Introduction to Permaculture", the names of two of his books. These applications were withdrawn in 2011. There has never been a trademark for the word *permaculture* in Australia.

Criticisms

General Criticisms

In 2011, Owen Hablutzel argued that "permaculture has yet to gain a large amount of specific mainstream scientific acceptance," and that "the sensitiveness to being perceived and accepted on scientific terms is motivated in part by a desire for permaculture to expand and become increasingly relevant." Bec-Hellouin permaculture farm engaged in a research program in partnership with INRA and AgroParisTech to collect scientific data.

In his books *Sustainable Freshwater Aquaculture* and *Farming in Ponds and Dams*, Nick Romanowski expresses the view that the presentation of aquaculture in Bill Mollison's books is unrealistic and misleading.

Agroforestry

Greg Williams argues that forests cannot be more productive than farmland because the net productivity of forests decline as they mature due to ecological succession. Proponents of permaculture respond that this is true only if one compares data between woodland forest and climax vegetation, but not when comparing farmland vegetation with woodland forest. For example, ecological succession generally results in a forest's productivity rising after its establishment only until it reaches the *woodland state* (67% tree cover), before declining until *full maturity*.

Weed Control

Weed control is the botanical component of pest control, which attempts to stop weeds, especially noxious or injurious weeds, from competing with domesticated plants and livestock. Many strategies have been developed in order to contain these plants.

The original strategy was manual removal including ploughing, which can cut the roots of weeds. More recent approaches include herbicides (chemical weed killers) and reducing stocks by burning and/or pulverizing seeds.

A plant is often termed a "weed" when it has one or more of the following characteristics:

- Little or no recognized value (as in medicinal, material, nutritional or energy)

- Rapid growth and/or ease of germination

- Competitive with crops for space, light, water and nutrients

The definition of a weed is completely context-dependent. To one person, one plant may be a weed, and to another person it may be a desirable plant. In one place, a plant may be viewed as a weed, whereas in another place, the same plant may be desirable.

Introduction

Weeds compete with productive crops or pasture, ultimately converting productive land into unusable scrub. Weeds can be poisonous, distasteful, produce burrs, thorns or otherwise interfere with the use and management of desirable plants by contaminating harvests or interfering with livestock.

Weeds compete with crops for space, nutrients, water and light. Smaller, slower growing seedlings are more susceptible than those that are larger and more vigorous. Onions are one of the most vulnerable, because they are slow to germinate and produce slender, upright stems. By contrast broad beans produce large seedlings and suffer far fewer effects other than during periods of water shortage at the crucial time when the pods are filling out. Transplanted crops raised in sterile soil or potting compost gain a head start over germinating weeds.

Weeds also vary in their competitive abilities and according to conditions and season. Tall-growing vigorous weeds such as fat hen (*Chenopodium album*) can have the most pronounced effects on adjacent crops, although seedlings of fat hen that appear in late summer produce only small plants. Chickweed (*Stellaria media*), a low growing plant, can happily co-exist with a tall crop during the summer, but plants that have overwintered will grow rapidly in early spring and may swamp crops such as onions or spring greens.

The presence of weeds does not necessarily mean that they are damaging a crop, especially during the early growth stages when both weeds and crops can grow without interference. However, as growth proceeds they each begin to require greater amounts of water and nutrients. Estimates suggest that weed and crop can co-exist harmoniously for around three weeks before competition becomes significant. One study found that after competition had started, the final yield of onion bulbs was reduced at almost 4% per day.

Perennial weeds with bulbils, such as lesser celandine and oxalis, or with persistent underground stems such as couch grass (*Agropyron repens*) or creeping buttercup (*Ranunculus repens*) store reserves of food, and are thus able to grow faster and with more vigour than their annual counterparts. Some perennials such as couch grass exude allelopathic chemicals that inhibit the growth of other nearby plants.

Weeds can also host pests and diseases that can spread to cultivated crops. Charlock and Shepherd's purse may carry clubroot, eelworm can be harboured by chickweed, fat hen and shepherd's purse, while the cucumber mosaic virus, which can devastate the cucurbit family, is carried by a range of different weeds including chickweed and groundsel.

Insect pests often do not attack weeds. However pests such as cutworms may first attack weeds then move on to cultivated crops.

Some plants are considered weeds by some farmers and crops by others. Charlock, a common weed in the southeastern US, are weeds according to row crop growers, but are valued by beekeepers, who seek out places where it blooms all winter, thus providing pollen for honeybees and other pollinators. Its bloom resists all but a very hard freeze, and recovers once the freeze ends.

Weed Propagation

Seeds

Annual and biennial weeds such as chickweed, annual meadow grass, shepherd's purse, groundsel, fat hen, cleaver, speedwell and hairy bittercress propagate themselves by seeding. Many produce huge numbers of seed several times a season, some all year round. Groundsel can produce 1000 seed, and can continue right through a mild winter, whilst Scentless Mayweedproduces over 30,000 seeds per plant. Not all of these will germinate at once, but over several seasons, lying dormant in the soil sometimes for years until exposed to light. Poppy seed can survive 80–100 years, dock 50 or more. There can be many thousands of seeds in a square foot or square metre of ground, thus and soil disturbance will produce a flush of fresh weed seedlings.

Subsurface/Surface

The most persistent perennials spread by underground creeping rhizomes that can regrow from a tiny fragment. These include couch grass, bindweed, ground elder, nettles, rosebay willow herb, Japanese knotweed, horsetail and bracken, as well as creeping thistle, whose tap roots can put out lateral roots. Other perennials put out runners that spread along the soil surface. As they creep they set down roots, enabling them to colonise bare ground with great rapidity. These include creeping buttercup and ground ivy. Yet another group of perennials propagate by stolons- stems that arch back into the ground to reroot. The most familiar of these is the bramble.

Methods

Weed control plans typically consist of many methods which are divided into biological, chemical, cultural, and physical/mechanical control.

Pesticide-free thermic weed control with a weed burner on a potato field in Dithmarschen

Physical/Mechanical Methods

Coverings

In domestic gardens, methods of weed control include covering an area of ground with a material that creates a hostile environment for weed growth, known as a *weed mat*.

Several layers of wet newspaper prevent light from reaching plants beneath, which kills them. Daily saturating the newspaper with water plant decomposition. After several weeks, all germinating weed seeds are dead.

In the case of black plastic, the greenhouse effect kills the plants. Although the black plastic sheet is effective at preventing weeds that it covers, it is difficult to achieve complete coverage. Eradicating persistent perennials may require the sheets to be left in place for at least two seasons.

Some plants are said to produce root exudates that suppress herbaceous weeds. *Tagetes minuta* is claimed to be effective against couch and ground elder, whilst a border of comfrey is also said to act as a barrier against the invasion of some weeds including couch. A 5–10 centimetres (2.0–3.9 in)} layer of wood chip mulch prevents most weeds from sprouting.

Gravel can serve as an inorganic mulch.

Irrigation is sometimes used as a weed control measure such as in the case of paddy fields to kill any plant other than the water-tolerant rice crop.

Manual Removal

Weeds are removed manually in large parts of India.

Many gardeners still remove weeds by manually pulling them out of the ground, making sure to include the roots that would otherwise allow them to resprout.

Hoeing off weed leaves and stems as soon as they appear can eventually weaken and kill perennials, although this will require persistence in the case of plants such as bindweed. Nettle infestations can be tackled by cutting back at least three times a year, repeated over a three-year period. Bramble can be dealt with in a similar way.

Tillage

Ploughing includes tilling of soil, intercultural ploughing and summer ploughing. Ploughing uproots weeds, causing them to die. In summer ploughing is done during deep summers. Summer ploughing also helps in killing pests.

Mechanical tilling can remove weeds around crop plants at various points in the growing process.

Thermal

Several thermal methods can control weeds.

Hot foam (foamstream) causes the cell walls to rupture, killing the plant. Weed burners heat up soil quickly and destroy superficial parts of the plants. Weed seeds are often heat resistant and even react with an increase of growth on dry heat.

Since the 19th century soil steam sterilization has been used to clean weeds completely from soil. Several research results confirm the high effectivness of humid heat against weeds and its seeds.

Soil solarization in some circumstances is very effective at eliminating weeds while maintaining grass. Planted grass tends to have a higher heat/humidity tolerance than unwanted weeds.

Seed Targeting

In 1998, the Australian Herbicide Resistance Initiative (AHRI), debuted. gathered fifteen scientists and technical staff members to conduct field surveys, collect seeds, test for resistance and study the biochemical and genetic mechanisms of resistance. A collaboration with DuPont led to a mandatory herbicide labeling program, in which each mode of action is clearly identified by a letter of the alphabet.

The key innovation of the AHRI approach has been to focus on weed seeds. Ryegrass seeds last only a few years in soil, so if farmers can prevent new seeds from arriving, the number of sprouts will shrink each year. Until the new approach farmers were unintentionally helping the seeds. Their combines loosen ryegrass seeds from their stalks and spread them over the fields. In the mid-1980s, a few farmers hitched covered trailers, called "chaff carts", behind their combines to catch the chaff and weed seeds. The collected material is then burned.

An alternative is to concentrate the seeds into a half-meter-wide strip called a windrow and burn the windrows after the harvest, destroying the seeds. Since 2003, windrow burning has been adopted by about 70% of farmers in Western Australia.

Yet another approach is the Harrington Seed Destructor, which is an adaptation of a coal pulverizing cage mill that uses steel bars whirling at up to 1500 rpm. It keeps all the organic material in the field and does not involve combustion, but kills 95% of seeds.

Cultural Methods

Stale Seed Bed

Another manual technique is the 'stale seed bed', which involves cultivating the soil, then leaving it fallow for a week or so. When the initial weeds sprout, the grower lightly hoes them away before planting the desired crop. However, even a freshly cleared bed is susceptible to airborne seed from elsewhere, as well as seed carried by passing animals on their fur, or from imported manure.

Buried Drip Irrigation

Buried drip irrigation involves burying drip tape in the subsurface near the planting bed, thereby limiting weeds access to water while also allowing crops to obtain moisture. It is most effective during dry periods.

Crop Rotation

Rotating crops with ones that kill weeds by choking them out, such as hemp, *Mucuna pruriens*, and other crops, can be a very effective method of weed control. It is a way to avoid the use of herbicides, and to gain the benefits of crop rotation.

Biological Methods

A biological weed control regiment can consist of biological control agents, bioherbicides, use of grazing animals, and protection of natural predators.

Animal Grazing

Companies using goats to control and eradicate leafy spurge, knapweed, and other toxic weeds have sprouted across the American West.

Chemical Methods

"Organic" Approaches

Weed control, circa 1930-40s

Organic weed control involves anything other than applying manufactured chemicals. Typically a combination of methods are used to achieve satisfactory control.

A mechanical weed control device: the diagonal weeder

Sulfur in some circumstances is accepted within British Soil Association standards.

Herbicides

The above described methods of weed control use no or very limited chemical inputs. They are preferred by organic gardeners or organic farmers.

However weed control can also be achieved by the use of herbicides. Selective herbicides kill certain targets while leaving the desired crop relatively unharmed. Some of these act by interfering with the growth of the weed and are often based on plant hormones. Herbicides are generally classified as follows:

- Contact herbicides destroy only plant tissue that contacts the herbicide. Generally, these are the fastest-acting herbicides. They are ineffective on perennial plants that can re-grow from roots or tubers.

- Systemic herbicides are foliar-applied and move through the plant where they destroy a greater amount of tissue. Glyphosate is currently the most used systemic herbicide.

- Soil-borne herbicides are applied to the soil and are taken up by the roots of the target plant.

- Pre-emergent herbicides are applied to the soil and prevent germination or early growth of weed seeds.

In agriculture large scale and systematic procedures are usually required, often by machines, such as large liquid herbicide 'floater' sprayers, or aerial application.

Bradley Method

Bradley Method of Bush Regeneration, which uses ecological processes to do much of the work. Perennial weeds also propagate by seeding; the airborne seed of the dandelion and the rose-bay willow herb parachute far and wide. Dandelion and dock also put down deep tap roots, which, although they do not spread underground, are able to regrow from any remaining piece left in the ground.

Hybrid

One method of maintaining the effectiveness of individual strategies is to combine them with others that work in complete different ways. Thus seed targeting has been combined with herbicides. In Australia seed management has been effectively combined with trifluralin and clethodim.

Resistance

Resistance occurs when a target adapts to circumvent a particular control strategy. It affects not only weed control,but antibiotics, insect control and other domains. In agriculture is mostly considered in reference to pesticides, but can defeat other strategies, e.g., when a target species becomes more drought tolerant via selection pressure.

Farming practices

Herbicide resistance recently became a critical problem as many Australian sheep farmers switched to exclusively growing wheat in their pastures in the 1970s. In wheat fields, introduced varieties of ryegrass, while good for grazing sheep, are intense competitors with wheat. Ryegrasses produce so many seeds that, if left unchecked, they can completely choke a field. Herbicides provided excellent control, while reducing soil disrupting because of less need to plough. Within little more than a decade, ryegrass and other weeds began to develop resistance. Australian farmers evolved again and began diversifying their techniques.

In 1983, patches of ryegrass had become immune to Hoegrass, a family of herbicides that inhibit an enzyme called acetyl coenzyme A carboxylase.

Ryegrass populations were large, and had substantial genetic diversity, because farmers had planted many varieties. Ryegrass is cross-pollinated by wind, so genes shuffle frequently. Farmers sprayed inexpensive Hoegrass year after year, creating selection pressure, but were diluting the herbicide in order to save money, increasing plants survival. Hoegrass was mostly replaced by a group of herbicides that block acetolactate synthase, again helped by poor application practices. Ryegrass evolved a kind of "cross-resistance" that allowed it to rapidly break down a variety of herbicides. Australian farmers lost four classes of herbicides in only a few years. As of 2013 only two herbicide classes, called Photosystem II and long-chain fatty acid inhibitors, had become the last hope.

Mechanical Weed Control

Mechanical weed control is any physical activity that inhibits unwanted plant growth. Mechanical, or manual, weed control techniques manage weed populations through physical methods that remove, injure, kill, or make the growing conditions unfavorable. Some of these methods cause direct damage to the weeds through complete removal or causing a lethal injury. Other techniques may alter the growing environment by eliminating light, increasing the temperature of the soil, or depriving the plant of carbon dioxide or oxygen. Mechanical control techniques can be either selective or non-selective. A selective method has very little impact on non-target plants where as a non-selective method affects the entire area that is being treated. If mechanical control methods are applied at the optimal time and intensity, some weed species may be controlled or even eradicated.

Mechanical Control Methods

Weed Pulling

Pulling methods uproot and remove the weed from the soil. Weed pulling can be used to control some shrubs, tree saplings, and herbaceous plants. Annuals and tap-rooted weeds tend to be very susceptible to pulling. Many species are able to re-sprout from root segments that are left in the soil. Therefore, the effectiveness of this method is dependent on the removal of as much of the root system as possible. Well established perennial weeds are much less effectively controlled because of the difficulty of removing all of the root system and perennating plant parts. Small herbaceous weeds may be pulled by hand but larger plants may require the use of puller tools like the Weed Wrench or the Root Talon. This technique has a little to no impact on neighboring, non-target plants and has a minimal effect on the growing environment. However, pulling is labor-intensive and time consuming making it a more suitable method to use for small weed infestations.

Mowing

Mowing methods cut or shred the above ground of the weed and can prevent and reduce seed populations as well as restrict the growth of weeds. Mowing can be a very successful control method for many annual weeds. Mowing is the most effective when it is performed before the weeds are able to set seed because it can reduce the number of flower stalks and prevent the spread of more seed. However, the biology of the weed must be considered before mowing. Some weed species may sprout with increased vigor after being mowed. Also, some species are able to re-sprout from stem or root segments that are left behind after mowing. Brush cutting and weed eating are also mowing techniques that reduce the biomass of the weeds. Repeatedly removing biomass causes reduced vigor in many weed species. This method is usually used in combination with other control methods such as burning or herbicide treatments.

Mulching

Mulch is a layer of material that is spread on the ground. Compared with some other methods of weed control, mulch is relatively simple and inexpensive. Mulching smothers the weeds by excluding light and providing a physical barrier to impede their emergence. Mulching is successful with most annual weeds, however, some perennial weeds are not affected. Mulches may be organic or synthetic. Organic mulches consist of plant by products such as: pine straw, wood chips, green waste, compost, leaves, and grass clippings. Synthetic mulches, also known as ground cover fabric, can be made from materials like polyethylene, polypropylene, or polyester. The effectiveness of mulching is mostly dependent on the material used. Organic and synthetic mulches may be used in combination with each other to increase the amount of weeds controlled.

Tillage

Tillage, also known as cultivation, is the turning over of the soil. This method is more often used in agricultural crops. Tillage can be performed on a small scale with tools such as small, hand pushed rotary tillers or on a large scale with tractor mounted plows. Tillage is able to control weeds because when the soil is overturned, the vegetative parts of the plants are damaged and the root systems are exposed causing desiccation. Generally, the younger the weed is, the more readily it can be controlled with tillage. To control mature perennial weeds, repeated tillage is necessary. By continually destroying new growth and damaging the root system, the weed's food stores are depleted until it can no longer re-sprout. Also, when the soil is overturned, the soil seed bank is disrupted which can cause dormant weed seeds to germinate in the absence of the previous competitors. These new weeds can also be controlled by continued tillage until the soil seed bank is depleted.

Soil Solarization

Soil solarization is a simple method of weed control that is accomplished by covering the soil with a layer of clear or black plastic. The plastic that is covering the ground traps heat energy from the sun and raises the temperature of the soil. Many weed seeds and vegetative propagules are not able to withstand the temperatures and are killed. For this method to be most effective, it should be implemented during the summer months and the soil should be moist. Also, cool season weeds are more susceptible to soil solarization than are warm season weeds. Using black plastic as a cover excludes light which can help to control plants that are growing whereas clear plastic has been shown to produce higher soil temperatures.

Fire

Burning and flaming can be economical and practical methods of weed control if used carefully. For most plants, fire causes the cell walls to rupture when they reach a temperature of 45 °C to 55 °C. Burning is commonly used to control weeds in forests, ditches, and roadsides. Burning can be used to remove accumulated vegetation by destroying the dry, matured plant matter as well as killing the green new growth. Buried weed seeds and plant propagules may also be destroyed during burning, however, dry seeds are much less susceptible to the increased temperature. Flaming is used on a smaller scale and includes the use of a propane torch with a fan tip. Flaming may be used to control weeds along fences and paved areas or places where the soil may be too wet to hoe, dig, or till. Flaming is most effective on young weeds that are less than two inches tall but repeated treatments may control tougher perennial weeds.

Flooding

Flooding is a method of control that requires the area being treated to be saturated at a depth of 15 to 30 cm for a period of 3 to 8 weeks. The saturation of the soil reduces the availability of oxygen to the plant roots thereby killing the weed. This method has been shown to be highly effective in controlling establish perennial weeds and may also suppress annual weeds by reducing the weed seed populations.

Effects of Mechanical Control on the Environment

Mechanical methods of weed control cause physical changes in the immediate environment that may cause positive or negative effects. The suppression of the targeted weeds will open niches in the environment and may also stimulate the growth of other weeds by decreasing their competition and making their environment more favorable. If the niches are not filled by a desirable plant, they will eventually be taken over by another weed. These weed control methods also effect the structure of the soil. The use of mulches can help decrease erosion, decrease water evaporation from the soil, as well as improve the soil structure by increasing the amount of organic matter. Tillage practices can help decrease compaction and aerate the soil. On the other hand, tillage has also been shown to decrease soil moisture, increase soil erosion and runoff, as well as decrease soil microbial populations. Solarization can cause changes in the biological, physical, and chemical properties of the soil. This can cause the soil to be an unfavorable environment for native species which may be beneficial or harmful.

Shifting Cultivation

Shifting cultivation is an agricultural system in which plots of land are cultivated temporarily, then abandoned and allowed to revert to their natural vegetation while the cultivator moves on to another plot. The period of cultivation is usually terminated when the soil shows signs of exhaustion or, more commonly, when the field is overrun by weeds. The length of time that a field is cultivated is usually shorter than the period over which the land is allowed to regenerate by lying fallow. This technique is often used in LEDCs (Less Economically Developed Countries) or LICs (Low Income Countries).

Slash-and-burn based shifting cultivation is a widespread historical practice in southeast Asia. Above is a satellite image of Sumatra and Borneo showing shift cultivation fires from October 2006.

Of these cultivators, many use a practice of slash-and-burn as one element of their farming cycle. Others employ land clearing without any burning, and some cultivators are purely migratory and do not use any cyclical method on a given plot. Sometimes no slashing at all is needed where regrowth is purely of grasses, an outcome not uncommon when soils are near exhaustion and need to lie fallow. In shifting agriculture, after two or three years of producing vegetable and grain crops on cleared land, the migrants abandon it for another plot. Trees and bushes are cleared by slashing, and the remaining vegetation is burnt. The ashes add potash to the soil. Then the seeds are sown after the rains

Advantages of Slash-and-Burn Method

Slash-and-burn is a very sustainable technique. It differs a lot from commercial farming, because once the trees are burned, there is very fertile fine ash that deposits along the humus, meaning that by the time the other fields are burned, the soil has time to reassemble nutriments, in order to make cultural activity possible. Although slash-and-burn is a very useful technique, there are other ways of fertilizing soil. By planting beans, the soil will regenerate much faster, due to the production of nitrogen in their roots.

Political Ecology of Shifting Cultivation

Shifting cultivation is a form of agriculture or a cultivation system, in which, at any particular point in time, a minority of 'fields' are in cultivation and a majority are in various stages of natural regrowth. Over time, fields are cultivated for a relatively short time, and allowed to recover, or are fallowed, for a relatively long time. Eventually a previously cultivated field will be cleared of the natural vegetation and planted in crops again. Fields in established and stable shifting cultivation systems are cultivated and fallowed cyclically.This type of farming is called jhumming in India.

Fallow fields are not unproductive. During the fallow period, shifting cultivators use the successive vegetation species widely for timber for fencing and construction, firewood, thatching, ropes, clothing, tools, carrying devices and medicines. It is common for fruit and nut trees to be planted in fallow fields to the extent that parts of some fallows are in fact orchards. Soil-enhancing shrub or tree species may be planted or protected from slashing or burning in fallows. Many of these species have been shown to fix nitrogen. Fallows commonly contain plants that attract birds and animals and are important for hunting. But perhaps most importantly, tree fallows protect soil against physical erosion and draw nutrients to the surface from deep in the soil profile.

The relationship between the time the land is cultivated and the time it is fallowed are critical to the stability of shifting cultivation systems. These parameters determine whether or not the shifting cultivation system as a whole suffers a net loss of nutrients over time. A system in which there is a net loss of nutrients with each cycle will eventually lead to a degradation of resources unless actions are taken to arrest the losses. In some cases soil can be irreversibly exhausted (including erosion as well as nutrient loss) in less than a decade.

The longer a field is cropped, the greater the loss of soil organic matter, cation-exchange-capacity and in nitrogen and phosphorus, the greater the increase in acidity, the more likely soil porosity and infiltration capacity is reduced and the greater the loss of seeds of naturally occurring plant species from soil seed banks. In a stable shifting cultivation system, the fallow is long enough for the natural vegetation to recover to the state that it was in before it was cleared, and for the soil to recover to the condition it was in before cropping began. During fallow periods soil temperatures are lower, wind and water erosion is much reduced, nutrient cycling becomes closed again, nutrients are extracted from the subsoil, soil fauna decreases, acidity is reduced, soil structure, texture and moisture characteristics improve and seed banks are replenished.

The secondary forests created by shifting cultivation are commonly richer in plant and animal resources useful to humans than primary forests, even though they are much less bio-diverse. Shifting cultivators view the forest as an agricultural landscape of fields at various stages in a regular cycle. People unused to living in forests cannot see the fields for the trees. Rather they perceive an apparently chaotic landscape in which trees are cut and burned randomly and so they characterise shifting cultivation as ephemeral or 'pre-agricultural', as 'primitive' and as a stage to be progressed beyond. Shifting agriculture is none of these things. Stable shifting cultivation systems are highly variable, closely adapted to micro-environments and are carefully managed by farmers during both the cropping and fallow stages. Shifting cultivators may possess a highly developed knowledge and understanding of their local environments and of the crops and native plant species they exploit. Complex and highly adaptive land tenure systems sometimes exist under shifting cultivation. Introduced crops for food and as cash have been skillfully integrated into some shifting cultivation systems.

Shifting Cultivation in Europe

Shifting cultivation was still being practised as a viable and stable form of agriculture in many parts of Europe and east into Siberia at the end of the 19th century and in some places well into the 20th century. In the Ruhr in the late 1860s a forest-field rotation system known as *Reutberg-*

wirtschaft was using a 16-year cycle of clearing, cropping and fallowing with trees to produce bark for tanneries, wood for charcoal and rye for flour (Darby 1956, 200). Swidden farming was practised in Siberia at least until the 1930s, using specially selected varieties of "swidden-rye" (Steensberg 1993, 98). In Eastern Europe and Northern Russia the main swidden crops were turnips, barley, flax, rye, wheat, oats, radishes and millet. Cropping periods were usually one year, but were extended to two or three years on very favourable soils. Fallow periods were between 20 and 40 years (Linnard 1970, 195). In Finland in 1949, Steensberg (1993, 111) observed the clearing and burning of a 60,000 square metres (15 acres) swidden 440 km north of Helsinki. Birch and pine trees had been cleared over a period of a year and the logs sold for cash. A fallow of alder (Alnus) was encouraged to improve soil conditions. After the burn, turnip was sown for sale and for cattle feed. Shifting cultivation was disappearing in this part of Finland because of a loss of agricultural labour to the industries of the towns. Steensberg (1993, 110-152) provides eye-witness descriptions of shifting cultivation being practised in Sweden in the 20th century, and in Estonia, Poland, the Caucasus, Serbia, Bosnia, Hungary, Switzerland, Austria and Germany in the 1930s to the 1950s.

That these agricultural practices survived from the Neolithic into the middle of the 20th century amidst the sweeping changes that occurred in Europe over that period, suggests they were adaptive and in themselves, were not massively destructive of the environments in which they were practised. This raises the question: if shifting cultivation did not lead to the disappearance of European forests, what did?

The earliest written accounts of forest destruction in Southern Europe begin around 1000 BC in the histories of Homer, Thucydides and Plato and in Strabo's Geography. Forests were exploited for ship building, and urban development, the manufacture of casks, pitch and charcoal, as well as being cleared for agriculture. The intensification of trade and as a result of warfare, increased the demand for ships which were manufactured completely from forest products. Although goat herding is singled out as an important cause of environmental degradation, a more important cause of forest destruction was the practice in some places of granting ownership rights to those who clear felled forests and brought the land into permanent cultivation. Evidence that circumstances other than agriculture were the major causes for forest destruction was the recovery of tree cover in many parts of the Roman empire from 400 BC to around 500 AD following the collapse of Roman economy and industry. Darby observes that by 400 AD "land that had once been tilled became derelict and overgrown" and quotes Lactantius who wrote that in many places "cultivated land became forest" (Darby 1956, 186). The other major cause of forest destruction in the Mediterranean environment with its hot dry summers were wild fires that became more common following human interference in the forests.

In Central and Northern Europe the use of stone tools and fire in agriculture is well established in the palynological and archaeological record from the Neolithic. Here, just as in Southern Europe, the demands of more intensive agriculture and the invention of the plough, trading, mining and smelting, tanning, building and construction in the growing towns and constant warfare, including the demands of naval shipbuilding, were more important forces behind the destruction of the forests than was shifting cultivation.

By the Middle Ages in Europe, large areas of forest were being cleared and converted into arable land in association with the development of feudal tenurial practices. From the 16th to the 18th centuries, the demands of iron smelters for charcoal, increasing industrial developments and the discovery and expansion of colonial empires as well as incessant warfare that increased the demand for shipping to levels never previously reached, all combined to deforest Europe. With the loss of the forest, so shifting cultivation became restricted to the peripheral places of Europe, where permanent agriculture was uneconomic, transport costs constrained logging or terrain prevented the use of draught animals or tractors. It has disappeared from even these refuges since 1945, as agriculture has become increasingly capital intensive, rural areas have become depopulated and the remnant European forests themselves have been revalued economically and socially.

Simple Societies, Shifting Cultivation and Environmental Change

Shifting cultivation in Indonesia. A new crop is sprouting through the burnt soil.

A growing body of palynological evidence finds that simple human societies brought about extensive changes to their environments before the establishment of any sort of state, feudal or capitalist, and before the development of large scale mining, smelting or shipbuilding industries. In these societies agriculture was the driving force in the economy and shifting cultivation was the most common type of agriculture practiced. By examining the relationships between social and economic change and agricultural change in these societies, insights can be gained on contemporary social and economic change and global environment change, and the place of shifting cultivation in those relationship.

As early as 1930 questions about relationships between the rise and fall of the Mayan civilization of the Yucatán Peninsula and shifting cultivation were raised and continue to be debated today. Archaeological evidence suggests the development of Mayan society and economy began around 250 AD. A mere 700 years later it reached its apogee, by which time the population may have reached 2,000,000 people. There followed a precipitous decline that left the great cities and ceremonial centres vacant and overgrown with jungle vegetation. The causes of this decline are uncertain; but warfare and the exhaustion of agricultural land are commonly cited (Meggers 1954; Dumond 1961; Turner 1974). More recent work suggests the Maya may have, in suitable places, developed irrigation systems and more intensive agricultural practices (Humphries 1993).

Similar paths appear to have been followed by Polynesian settlers in New Zealand and the Pacific Islands, who within 500 years of their arrival around 1100 AD turned substantial areas from forest into scrub and fern and in the process caused the elimination of numerous species of birds and animals (Kirch and Hunt 1997). In the restricted environments of the Pacific islands, including Fiji and Hawaii, early extensive erosion and change of vegetation is presumed to have been caused by shifting cultivation on slopes. Soils washed from slopes were deposited in valley bottoms as a rich, swampy alluvium. These new environments were then exploited to develop intensive, irrigated fields. The change from shifting cultivation to intensive irrigated fields occurred in association with a rapid growth in population and the development of elaborate and highly stratified chiefdoms (Kirch 1984). In the larger, temperate latitude, islands of New Zealand the presumed course of events took a different path. There the stimulus for population growth was the hunting of large birds to extinction, during which time forests in drier areas were destroyed by burning, followed the development of intensive agriculture in favorable environments, based mainly on sweet potato (Ipomoea batatas) and a reliance on the gathering of two main wild plant species in less favorable environments. These changes, as in the smaller islands, were accompanied by population growth, the competition for the occupation of the best environments, complexity in social organization, and endemic warfare (Anderson 1997).

The record of humanly induced changes in environments is longer in New Guinea than in most places. Agricultural activities probably began 5,000 to 9,000 years ago. However, the most spectacular changes, in both societies and environments, are believed to have occurred in the central highlands of the island within the last 1,000 years, in association with the introduction of a crop new to New Guinea, the sweet potato (Golson 1982a; 1982b). One of the most striking signals of the relatively recent intensification of agriculture is the sudden increase in sedimentation rates in small lakes.

The root question posed by these and the numerous other examples that could be cited of simple societies that have intensified their agricultural systems in association with increases in population and social complexity is not whether or how shifting cultivation was responsible for the extensive changes to landscapes and environments. Rather it is why simple societies of shifting cultivators in the tropical forest of Yucatán, or the highlands of New Guinea, began to grow in numbers and to develop stratified and sometimes complex social hierarchies?

At first sight, the greatest stimulus to the intensification of a shifting cultivation system is a growth in population. If no other changes occur within the system, for each extra person to be fed from the system, a small extra amount of land must be cultivated. The total amount of land available is the land being presently cropped and all of the land in fallow. If the area occupied by the system is not expanded into previously unused land, then either the cropping period must be extended or the fallow period shortened.

At least two problems exist with the population growth hypothesis. First, population growth in most pre-industrial shifting cultivator societies has been shown to be very low over the long term. Second, no human societies are known where people work only to eat. People engage in social relations with each other and agricultural produce is used in the conduct of these relationships.

These relationships are the focus of two attempts to understand the nexus between human societies and their environments, one an explanation of a particular situation and the other a general exploration of the problem.

1. Feedback Loops

In a study of the Duna in the Southern Highlands of New Guinea, a group in the process of moving from shifting cultivation into permanent field agriculture post sweet potato, Modjeska (1982) argued for the development of two "self amplifying feed back loops" of ecological and social causation. The trigger to the changes was very slow population growth and the slow expansion of agriculture to meet the demands of this growth. This set in motion the first feedback loop, the "use-value" loop. As more forest was cleared there was a decline in wild food resources and protein produced from hunting, which was substituted for by an increase in domestic pig raising. An increase in domestic pigs required a further expansion in agriculture. The greater protein available from the larger number of pigs increased human fertility and survival rates and resulted in faster population growth.

The outcome of the operation of the two loops, one bringing about ecological change and the other social and economic change, is an expanding and intensifying agricultural system, the conversion of forest to grassland, a population growing at an increasing rate and expanding geographically and a society that is increasing in complexity and stratification.

2. Resources are Cultural Appraisals

The second attempt to explain the relationships between simple agricultural societies and their environments is that of Ellen (1982, 252-270). Ellen does not attempt to separate use-values from social production. He argues that almost all of the materials required by humans to live (with perhaps the exception of air) are obtained through social relations of production and that these relations proliferate and are modified in numerous ways. The values that humans attribute to items produced from the environment arise out of cultural arrangements and not from the objects themselves, a restatement of Carl Sauer's dictum that "resources are cultural appraisals". Humans frequently translate actual objects into culturally conceived forms, an example being the translation by the Duna of the pig into an item of compensation and redemption. As a result, two fundamental processes underlie the ecology of human social systems: First, the obtaining of materials from the environment and their alteration and circulation through social relations, and second, giving the material a value which will affect how important it is to obtain it, circulate it or alter it. Environmental pressures are thus mediated through social relations.

Transitions in ecological systems and in social systems do not proceed at the same rate. The rate of phylogenetic change is determined mainly by natural selection and partly by human interference and adaptation, such as for example, the domestication of a wild species. Humans however have the ability to learn and to communicate their knowledge to each other and across generations. If most social systems have the tendency to increase in complexity they will, sooner or later, come into conflict with, or into "contradiction" (Friedman 1979, 1982) with their environments. What happens around the point of "contradiction" will determine the extent of the environmental degradation that will occur. Of particular importance is the ability of the society to change, to invent or to innovate technologically and sociologically, in order to overcome the "contradiction" without incurring continuing environmental degradation, or social disintegration.

An economic study of what occurs at the points of conflict with specific reference to shifting cultivation is that of Esther Boserup (1965). Boserup argues that low intensity farming, extensive shifting cultivation for example, has lower labor costs than more intensive farming systems. This assertion remains controversial. She also argues that given a choice, a human group will always choose the technique which has the lowest absolute labor cost rather than the highest yield. But at the point of conflict, yields will have become unsatisfactory. Boserup argues, contra Malthus, that rather than population always overwhelming resources, that humans will invent a new agricultural technique or adopt an existing innovation that will boost yields and that is adapted to the new environmental conditions created by the degradation which has occurred already, even though they will pay for the increases in higher labor costs. Examples of such changes are the adoption of new higher yielding crops, the exchanging of a digging stick for a hoe, or a hoe for a plough, or the development of irrigation systems. The controversy over Boserup's proposal is in part over whether intensive systems are more costly in labor terms, and whether humans will bring about change in their agricultural systems before environmental degradation forces them to.

Shifting Cultivation in the Contemporary World and Global Environmental Change

Contemporary Shifting Cultivation Practice

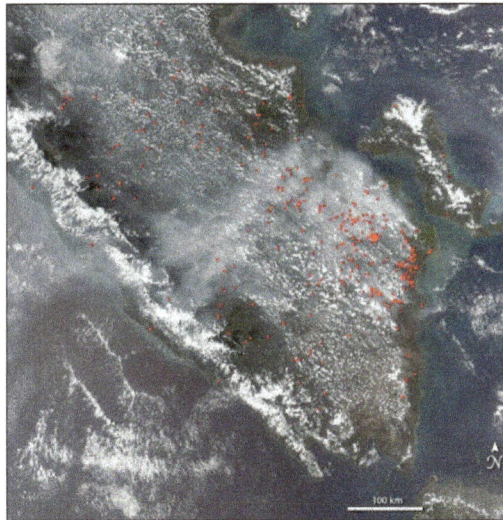

Sumatra, Indonesia

Rio Xingu, Brazil

Santa Cruz, Bolivia

The estimated rate of deforestation in Southeast Asia in 1990 was 34,000 km² per year (FAO 1990, quoted in Potter 1993). In Indonesia alone it was estimated 13,100 km² per year were being lost, 3,680 km² per year from Sumatra and 3,770 km² from Kalimantan, of which 1,440 km² were due to the fires of 1982 to 1983. Since those estimates were made huge fires have ravaged Indonesian forests during the 1997 to 1998 El Niño associated drought.

Kasempa , Zambia

Interdisciplinary Project

Shifting cultivation used to be the backbone of smallholder agriculture throughout the tropics, but today it is abandoned in many places in favor of large scale cash crop production – e.g. for biofuels. The extent of these changes is not well documented because shifting cultivation land rarely appears on official maps and census data seldom identifies shifting cultivators. Moreover, the consequences of these changes for livelihoods (e.g. food security) are not well known. The aim of this project is to analyze the extent and consequences of change in shifting cultivation by combining meta-analyses of existing studies and census data with case studies in selected areas. This interdisciplinary project focuses on:

1. Trends in change in shifting cultivation landscapes and demography; and

2. Changes in livelihoods due to these changes.

The project will compile data for eight countries (Mexico, Brazil, Laos, Vietnam, Malaysia, Thailand, Zambia and Tanzania) and the outcome is expected to be relevant to planning and policy-making on land and forest management.

Shifting cultivation was assessed by the FAO to be one a causes of deforestation while logging was not. The apparent discrimination against shifting cultivators caused a confrontation between FAO and environmental groups, who saw the FAO supporting commercial logging interests against the rights of indigenous people (Potter 1993, 108). Other independent studies of the problem note that despite lack of government control over forests and the dominance of a political elite in the logging industry, the causes of deforestation are more complex. The loggers have provided paid employment to former subsistence farmers. One of the outcomes of cash incomes has been rapid population growth among indigenous groups of former shifting cultivators that has placed pressure on their traditional long fallow farming systems. Many farmers have taken advantage of the improved road access to urban areas by planting cash crops, such as rubber or pepper as noted above. Increased cash incomes often are spent on chain saws, which have enabled larger areas to be cleared for cultivation. Fallow periods have been reduced and cropping periods extended. Serious poverty elsewhere in the country has brought thousands of land hungry settlers into the cut over forests along the logging roads. The settlers practice what appears to be shifting cultivation but which is in fact a one-cycle slash and burn followed by continuous cropping, with no intention to long fallow. Clearing of trees and the permanent cultivation of fragile soils in a tropical environment with little attempt to replace lost nutrients may cause rapid degradation of the fragile soils.

The loss of forest in Indonesia, Thailand, and the Philippines during the 1990s was preceded by major ecosystem disruptions in Vietnam, Laos and Cambodia in the 1970s and 1980s caused by warfare. Forests were sprayed with defoliants, thousands of rural forest dwelling people uproots from their homes and moved and roads driven into previously isolated areas. The loss of the tropical forests of Southeast Asia is the particular outcome of the general possible outcomes described by Ellen when small local ecological and social systems become part of larger system. When the previous relatively stable ecological relationships are destabilized, degradation can occur rapidly. Similar descriptions of the loss of forest and destruction of fragile ecosystems could be provided from the Amazon Basin, by large scale state sponsored colonization forest land (Becker 1995, 61) or from the Central Africa where what endemic armed conflict is destabilizing rural settlement and farming communities on a massive scale.

Comparison with Other Ecological Phenomena

In the tropical developing world, shifting cultivation in its many diverse forms, remains a pervasive practice. Shifting cultivation was one of the very first forms of agriculture practiced by humans and its survival into the modern world suggests that it is a flexible and highly adaptive means of production. However, it is also a grossly misunderstood practice. Many casual observers cannot see past the clearing and burning of standing forest and do not perceive often ecologically stable cycles of cropping and fallowing. Nevertheless, shifting cultivation systems are particularly susceptible to rapid increases in population and to economic and social change in the larger world around them. The blame for the destruction of forest resources is often laid on shifting cultivators. But the forces bringing about the rapid loss of tropical forests at the end of the 20th century are the same forces that led to the destruction of the forests of Europe, urbanization, industrialization, increased affluence, populational growth and geographical expansion and the application the latest technology to extract ever more resources from the environment in pursuit of wealth and political power by competing groups. However we must know that those who practice Agriculture are at the receiving end of the social stratum.

Studies of small, isolated and pre-capitalist groups and their relationships with their environments suggests that the roots of the contemporary problem lie deep in human behavioral patterns, for even in these simple societies, competition and conflict can be identified as the main force driving them into contradiction with their environments.

Alternative Practice in the Pre-Columbian Amazon Basin

Slash-and-char, as opposed to slash-and-burn, may create self-perpetuating soil fertility that supports sedentary agriculture, but the society so sustained may still be overturned, as above.

Forest Gardening

Robert Hart's forest garden in Shropshire

Forest gardening is a low maintenance sustainable plant-based food production and agroforestry system based on woodland ecosystems, incorporating fruit and nut trees, shrubs, herbs, vines and perennial vegetables which have yields directly useful to humans. Making use of companion planting, these can be intermixed to grow in a succession of layers, to build a woodland habitat.

Forest gardening is a prehistoric method of securing food in tropical areas. In the 1980s, Robert Hart coined the term "forest gardening" after adapting the principles and applying them to temperate climates.

History

Forest gardens are probably the world's oldest form of land use and most resilient agroecosystem. They originated in prehistoric times along jungle-clad river banks and in the wet foothills of monsoon regions. In the gradual process of families improving their immediate environment, useful tree and vine species were identified, protected and improved whilst undesirable species were eliminated. Eventually superior foreign species were selected and incorporated into the gardens.

Forest gardens are still common in the tropics and known by various names such as: *home gardens* in Kerala in South India, Nepal, Zambia, Zimbabwe and Tanzania; *Kandyan forest gardens* in Sri Lanka; *huertos familiares*, the "family orchards" of Mexico; and *pekarangan*, the gardens of "complete design", in Java. These are also called agroforests and, where the wood components are short-statured, the term shrub garden is employed. Forest gardens have been shown to be a significant source of income and food security for local populations.

Robert Hart adapted forest gardening for the United Kingdom's temperate climate during the 1980s. His theories were later developed by Martin Crawford from the Agroforestry Research Trust and various permaculturalists such as Graham Bell, Patrick Whitefield, Dave Jacke and Geoff Lawton.

In Tropical Climates

Forest gardens, or home gardens, are common in the tropics, using intercropping to cultivate trees, crops, and livestock on the same land. In Kerala in south India as well as in northeastern India, the home garden is the most common form of land use and is also found in Indonesia. One example combines coconut, black pepper, cocoa and pineapple. These gardens exemplify polyculture, and conserve much crop genetic diversity and heirloom plants that are not found in monocultures. Forest gardens have been loosely compared to the religious concept of the Garden of Eden.

Americas

The BBC's *Unnatural Histories* claimed that the Amazon rainforest, rather than being a pristine wilderness, has been shaped by humans for at least 11,000 years through practices such as forest gardening and *terra preta*. This was also explored in the bestselling book *1491* by author Charles C. Mann. Since the 1970s, numerous geoglyphs have also been discovered on deforested land in the Amazon rainforest, furthering the evidence about Pre-Columbian civilizations.

On the Yucatán Peninsula, much of the Maya food supply was grown in "orchard-gardens", known as *pet kot*. The system takes its name from the low wall of stones (*pet* meaning circular and *kot* wall of loose stones) that characteristically surrounds the gardens.

Africa

In many African countries, for example Zambia, Zimbabwe, Ethiopia and Tanzania, gardens are widespread in rural, periurban and urban areas and they play an essential role in establishing food security. Most well known are the Chaga or Chagga gardens on the slopes of Mt. Kilimanjaro in Tanzania. These are an excellent example of an agroforestry system. In many countries, women are the main actors in home gardening and food is mainly produced for subsistence. In North-Africa, oasis layered gardening with palm trees, fruit trees and vegetables is a traditional type of forest garden.

Nepal

In Nepal, the *Ghar Bagaincha*, literally "home garden", refers to the traditional land use system around a homestead, where several species of plants are grown and maintained by household members and their products are primarily intended for the family consumption (Shrestha et al., 2002). The term "home garden" is often considered synonymous to the kitchen garden. However, they differ in terms of function, size, diversity, composition and features (Sthapit et al., 2006). In Nepal, 72% of households have home gardens of an area 2–11% of the total land holdings (Gautam et al., 2004). Because of their small size, the government has never identified home gardens as an important unit of food production and they thereby remain neglected from research and development. However, at the household level the system is very important as it is an important source of quality food and nutrition for the rural poor and, therefore, are important contributors to the household food security and livelihoods of farming communities in Nepal. The gardens are typically cultivated with a mixture of annual and perennial plants that can be harvested on a daily or seasonal basis. Biodiversity that has an immediate value is maintained in home gardens as women and children have easy access to preferred food. Home gardens, with their intensive and multiple uses, provide a safety net for households when food is scarce. These gardens are not only important sources of food, fodder, fuel, medicines, spices, herbs, flowers, construction materials and income in many countries, they are also important for the in situ conservation of a wide range of unique genetic resources for food and agriculture (Subedi et al., 2004). Many uncultivated, as well as neglected and underutilised species could make an important contribution to the dietary diversity of local communities (Gautam et al., 2004).

In addition to supplementing diet in times of difficulty, home gardens promote whole-family and whole-community involvement in the process of providing food. Children, the elderly, and those caring for them can participate in this infield agriculture, incorporating it with other household tasks and scheduling. This tradition has existed in many cultures around the world for thousands of years.

In Mediterranean Climates

The mediterranean climate has long, hot, rainless summers and relatively short, cool, rainy winters (Köppen climate classification *Csa*). Its climate conditions are highly variable within an area and modified locally by altitude, latitude, and the proximity to the Mediterranean. In the 1950s the Forest Research Department of the Ministry of Agriculture founded a botanical forest garden in the Sharon region in Israel, the Ilanot Forest. As the only one of its kind in Israel, it harbours more than seven hundred and fifty species of trees from locations all over the world, including the Japanese sago palm cycas revoluta, fig trees (ficus glomerata), stone pine trees (pinus pinea) that produce tasty pine nuts and adds to the biodiversity of Israel.

In Temperate Climates

Robert Hart, forest gardening pioneer

Robert Hart coined the term "forest gardening" during the 1980s. Hart began farming at Wenlock Edge in Shropshire with the intention of providing a healthy and therapeutic environment for himself and his brother Lacon. Starting as relatively conventional smallholders, Hart soon discovered that maintaining large annual vegetable beds, rearing livestock and taking care of an orchard were tasks beyond their strength. However, a small bed of perennial vegetables and herbs he planted was looking after itself with little intervention.

Following Hart's adoption of a raw vegan diet for health and personal reasons, he replaced his farm animals with plants. The three main products from a forest garden are fruit, nuts and green leafy vegetables. He created a model forest garden from a 0.12 acre (500 m²) orchard on his farm and intended naming his gardening method *ecological horticulture* or *ecocultivation*. Hart later dropped these terms once he became aware that *agroforestry* and *forest gardens* were already being used to describe similar systems in other parts of the world. He was inspired by the forest farming methods of Toyohiko Kagawa and James Sholto Douglas, and the productivity of the Keralan home gardens as Hart explains:

From the agroforestry point of view, perhaps the world's most advanced country is the Indian state of Kerala, which boasts no fewer than three and a half million forest gardens…As an example of the extraordinary intensivity of cultivation of some forest gardens, one plot of only 0.12 hectares (0.30 acres) was found by a study group to have twenty-three young coconut palms, twelve cloves, fifty-six bananas, and forty-nine pineapples, with thirty pepper vines trained up its trees. In addition, the small holder grew fodder for his house-cow.

Seven-Layer System

1. CANOPY (LARGE FRUIT & NUT TREES)
2. LOW TREE LAYER (DWARF FRUIT TREES)
3. SHRUB LAYER (CURRANTS & BERRIES)
4. HERBACEOUS (COMFREYS, BEETS, HERBS)
5. RHIZOSPHERE (ROOT VEGETABLES)
6. SOIL SURFACE (GROUND COVER, EG, STRAWBERRY, ETC)
7. VERTICAL LAYER (CLIMBERS, VINES)

THE FOREST GARDEN: A SEVEN LEVEL BENEFICIAL GUILD

The seven layers of the forest garden

Robert Hart pioneered a system based on the observation that the natural forest can be divided into distinct levels. He used intercropping to develop an existing small orchard of apples and pears into an edible polyculture landscape consisting of the following layers:

1. 'Canopy layer' consisting of the original mature fruit trees.

2. 'Low-tree layer' of smaller nut and fruit trees on dwarfing root stocks.

3. 'Shrub layer' of fruit bushes such as currants and berries.

4. 'Herbaceous layer' of perennial vegetables and herbs.

5. 'Rhizosphere' or 'underground' dimension of plants grown for their roots and tubers.

6. 'Ground cover layer' of edible plants that spread horizontally.

7. 'Vertical layer' o f vines and climbers.

A key component of the seven-layer system was the plants he selected. Most of the traditional vegetable crops grown today, such as carrots, are sun loving plants not well selected for the more shady forest garden system. Hart favoured shade tolerant perennial vegetables.

Further Development

The Agroforestry Research Trust (ART), managed by Martin Crawford, runs experimental forest gardening projects on a number of plots in Devon, United Kingdom. Crawford describes a forest garden as a low-maintenance way of sustainably producing food and other household products.

Ken Fern had the idea that for a successful temperate forest garden a wider range of edible shade tolerant plants would need to be used. To this end, Fern created the organisation Plants for a Future (PFAF) which compiled a plant database suitable for such a system. Fern used the term *woodland gardening*, rather than forest gardening, in his book *Plants for a Future*.

The Movement for Compassionate Living (MCL) promote forest gardening and other types of vegan organic gardening to meet society's needs for food and natural resources. Kathleen Jannaway, the founder of MCL, wrote a book outlining a sustainable vegan future called *Abundant Living in the Coming Age of the Tree* in 1991. In 2009, the MCL provided a grant of £1,000 to the Bangor Forest Garden project in Gwynedd, North West Wales.

Kevin Bradley coined the phrase "Edible Forest" in the 1980s as the name of his nursery, garden, and orchard on 5 acres in the frigid zone 3 pine forests of northern Wisconsin. Among 3 options, he chose "Edible Forest" because it "evokes at once an ethereal, spiritual, and magical image", of Disney- like "Forest of No Return"; of the biblical "Garden of Eden". This image was perfectly in line with his ongoing experiment begun in 1985 in what he calls a closed loop human environment, combining multi- story tree and field crop "garden/orchards" for maximum beauty and use of space, someday to be very useful in an ever shrinking world. "The name, at the same time, with its irrational first impression (of course we can't eat a forest), forces the mind to think, if just a little bit, about its inference and thus sticks in our memories". It appeared from Bradley's research that the two words had, prior to the 80's, never been put together before as a noun phrase but which by today, after more than two decades of Bradley's "Edible Forest Nursery" and the 2005 text by Jacke and Toensmeirer's- "Edible Forest Gardens", has grown into a movement and little "Edible Forests" all over the world.

In 2005, Dave Jacke and Eric Toensmeier's two-volume *Edible Forest Gardens* provided a deeply researched reference focused on North American forest gardening climates, habitats, and species. The book attempts to ground forest gardening deeply in ecological science. The Apios Institute wiki grew out of their work, and seeks to document and share the experience of people around the world working with the species in polycultures.

Permaculture

Bill Mollison, who coined the term *permaculture*, visited Robert Hart at his forest garden in Wenlock Edge in October 1990. Hart's seven-layer system has since been adopted as a common permaculture design element.

Numerous permaculturalists are proponents of forest gardens, or food forests, such as Graham Bell, Patrick Whitefield, Dave Jacke, Eric Toensmeier and Geoff Lawton. Bell started building his forest garden in 1991 and wrote the book *The Permaculture Garden* in 1995, Whitefield wrote the book *How to Make a Forest Garden* in 2002, Jacke and Toensmeier co-authored the two volume book set *Edible Forest Gardening* in 2005, and Lawton presented the film *Establishing a Food*

Forest in 2008.

Austrian Sepp Holzer practices "Holzer Permaculture" on his *Krameterhof* farm, at varying altitudes ranging from 1,100 to 1,500 metres above sea level. His designs create micro-climates with rocks, ponds and living wind barriers, enabling the cultivation of a variety of fruit trees, vegetables and flowers in a region that averages 4 °C, and with temperatures as low as -20 °C in the winter.

Projects

El Pilar on the Belize-Guatemala border features a forest garden to demonstrate traditional Maya agricultural practices. A further 1-acre model forest garden, called Känan K'aax (meaning well-tended garden in Mayan), is being funded by the National Geographic Society and developed at Santa Familia Primary School in Cayo.

In the United States the largest known food forest on public land is believed to be the 7-acre Beacon Food Forest in Seattle, WA. Other forest garden projects include those at the Central Rocky Mountain Permaculture Institute in Basalt, Colorado and Montview Neighborhood farm in Northampton, Massachusetts.

In Canada food forester Richard Walker has been developing and maintaining food forests in the province of British Columbia for over 30 years. He developed a 3-acre food forest that when at maturity provided raw materials for a nursery and herbalism business as well as food for his family. The Living Centre have developed various forest garden projects in Ontario.

In the United Kingdom, other than those run by the Agroforestry Research Trust (ART), there are numerous forest garden projects such as the Bangor Forest Garden in Gwynedd, North West Wales. Martin Crawford from ART administers the Forest Garden Network, an informal network of people and organisations around the world who are cultivating their own forest gardens.

References

- Dufour, Rex (July 2015). Tipsheet: Crop Rotation in Organic Farming Systems (Report). National Center for Appropriate Technology. Retrieved May 4, 2016.

- Baldwin, Keith R. (June 2006). Crop Rotations on Organic Farms (PDF) (Report). Center for Environmental Farming Systems. Retrieved May 4, 2016.

- Johnson, Sue Ellen; Charles L. Mohler, (2009). Crop Rotation on Organic Farms: A Planning Manual, NRAES 177. Ithica, NY: National Resource, Agriculture, and Engineering Services (NRAES). ISBN 978-1-933395-21-0.

- Coleman, Pamela (November 2012). Guide for Organic Crop Producers (PDF) (Report). National Organic Program. Retrieved May 4, 2016.

- Lamb, John; Craig Sheaffer & Kristine Moncada (2010). "Chapter 4 Soil Fertility". Risk Management Guide for Organic Producers (Report). University of Minnesota.

- Gegner, Lance; George Kuepper (August 2004). "Organic Crop Production Overview". National Center for Appropriate Technology. Retrieved May 4, 2016.

- Victoria, Reynaldo (2012). "The Benefits of Soil Carbon". Risk Management Guide for Organic Producers (Report). United Nations Environment Programme.

Various Practices of Sustainable Agriculture

Due to environmental concerns and depletion of natural resources, it is extremely important to practice sustainable agriculture globally. The following chapter will give a glimpse of the popular practices that are prevalent for example, wind break, compost, no-till farming, afforestation, etc. which will help readers broaden their spectrum of knowledge.

Intercropping

Intercropping is a multiple cropping practice involving growing two or more crops in proximity. The most common goal of intercropping is to produce a greater yield on a given piece of land by making use of resources that would otherwise not be utilized by a single crop. Careful planning is required, taking into account the soil, climate, crops, and varieties. It is particularly important not to have crops competing with each other for physical space, nutrients, water, or sunlight. Examples of intercropping strategies are planting a deep-rooted crop with a shallow-rooted crop, or planting a tall crop with a shorter crop that requires partial shade. Inga alley cropping has been proposed as an alternative to the ecological destruction of slash-and-burn farming.

When crops are carefully selected, other agronomic benefits are also achieved. Lodging-prone plants, those that are prone to tip over in wind or heavy rain, may be given structural support by their companion crop. Creepers can also benefit from structural support. Some plants are used to suppress weeds or provide nutrients. Delicate or light-sensitive plants may be given shade or protection, or otherwise wasted space can be utilized. An example is the tropical multi-tier system where coconut occupies the upper tier, banana the middle tier, and pineapple, ginger, or leguminous fodder, medicinal or aromatic plants occupy the lowest tier.

Intercropping of compatible plants also encourages biodiversity, by providing a habitat for a variety of insects and soil organisms that would not be present in a single-crop environment. This in turn can help limit outbreaks of crop pests by increasing predator biodiversity. Additionally, reducing the homogeneity of the crop increases the barriers against biological dispersal of pest organisms through the crop.

The degree of spatial and temporal overlap in the two crops can vary somewhat, but both requirements must be met for a cropping system to be an intercrop. Numerous types of intercropping, all of which vary the temporal and spatial mixture to some degree, have been identified. These are some of the more significant types:

- Mixed intercropping, as the name implies, is the most basic form in which the component crops are totally mixed in the available space.

- Row cropping involves the component crops arranged in alternate rows. Variations include alley cropping, where crops are grown in between rows of trees, and strip cropping, where multiple rows, or a strip, of one crop are alternated with multiple rows of another crop. A new version of this is to intercrop rows of solar photovoltaic modules with agriculture crops. This practice is called agrivoltaics.

- Temporal intercropping uses the practice of sowing a fast-growing crop with a slow-growing crop, so that the fast-growing crop is harvested before the slow-growing crop starts to mature.

- Further temporal separation is found in relay cropping, where the second crop is sown during the growth, often near the onset of reproductive development or fruiting, of the first crop, so that the first crop is harvested to make room for the full development of the second.

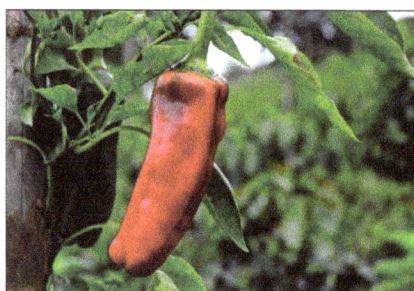

Chili pepper intercropped with coffee in Colombia's southwestern Cauca Department

Coconut and *Tagetes erecta*, a multilayer cropping in India

Windbreak

Aerial view of field windbreaks in North Dakota

A windbreak or shelterbelt is a plantation usually made up of one or more rows of trees or shrubs planted in such a manner as to provide shelter from the wind and to protect soil from erosion.

They are commonly planted around the edges of fields on farms. If designed properly, windbreaks around a home can reduce the cost of heating and cooling and save energy. Windbreaks are also planted to help keep snow from drifting onto roadways and even yards. Other benefits include providing habitat for wildlife and in some regions the trees are harvested for wood products.

Windbreaks and intercropping can be combined in a farming practice referred to as alleycropping. Fields are planted in rows of different crops surrounded by rows of trees. These trees provide fruit, wood, or protect the crops from the wind. Alley cropping has been particularly successful in India, Africa, and Brazil, where coffee growers have combined farming and forestry.

A further use for a shelterbelt is to screen a farm from a main road or motorway. This improves the farm landscape by reducing the visual incursion of the motorway, mitigating noise from the traffic and providing a safe barrier between farm animals and the road.

The term "windbreak" is also used to describe an article of clothing worn to prevent wind chill. Americans tend to use the term "windbreaker" whereas Europeans favor the term "windbreak".

Fences called "windbreaks" are also used. Normally made from cotton, nylon, canvas, and recycled sails, windbreaks tend to have three or more panels held in place with poles that slide into pockets sewn into the panel. The poles are then hammered into the ground and a windbreak is formed. Windbreaks or "wind fences" are used to reduce wind speeds over erodible areas such as open fields, industrial stockpiles, and dusty industrial operations. As erosion is proportional to wind speed cubed a reduction of wind speed of 1/2 (for example) will reduce erosion by over 80%.

Windbreak Aerodynamics

An East German windbreak promotion poster, 1952

In essence, when the wind encounters a porous obstacle such as a windbreak or shelterbelt, air pressure increases (loosely speaking, air *piles up*) on the windward side and (conversely) air pressure decreases on the leeward side. As a result, the airstream approaching the barrier is retarded, and a proportion of it is displaced up and over the barrier, resulting in a *jet* of higher wind speed aloft. The remainder of the impinging airstream, having been retarded in its approach, now circulates through the barrier to its downstream edge, pushed along by the decrease in pressure across the shelterbelt's width; emerging on the downwind side, that airstream is now further retarded by an adverse pressure gradient, because in the lee of the barrier, with increasing downwind distance air pressure *recovers* again to the ambient level. The result is that minimum wind speed occurs not at or within the windbreak, nor at its downwind edge, but further downwind - nominally, at a distance of about 3 to 5 times the windbreak height H. Beyond that point wind speed recovers, aided by downward momentum transport from the overlying, faster-moving stream. From the perspective of the Reynolds-averaged Navier–Stokes equations these effects can be understood as resulting from the loss of momentum caused by the drag of leaves and branches and would be represented by the body force f_i (a distributed momentum sink).

Not only is the mean (average) wind speed reduced in the lee of the shelter, the wind is also less gusty, for turbulent wind fluctuations are also damped. As a result, turbulent vertical mixing is weaker in the lee of the barrier than it is upwind, and interesting secondary microclimatic effects result. For instance, by day sensible heat rising from the ground due to the absorption of sunlight is mixed upward less efficiently in the lee of a windbreak, with the result that air temperature near ground is somewhat higher in the lee than on the windward side. Of course this effect is attenuated with increasing downwind distance and indeed, beyond about $8H$ downstream a zone may exist that is actually *cooler* than upwind.

Compost

A community-level composting plant in a rural area in Germany

Compost is organic matter that has been decomposed and recycled as a fertilizer and soil amendment. Compost is a key ingredient in organic farming.

At the simplest level, the process of composting simply requires making a heap of wetted organic matter known as green waste (leaves, food waste) and waiting for the materials to break down into humus after a period of weeks or months. Modern, methodical composting is a multi-step, closely monitored process with measured inputs of water, air, and carbon- and nitrogen-rich materials.

The decomposition process is aided by shredding the plant matter, adding water and ensuring proper aeration by regularly turning the mixture. Worms and fungi further break up the material. Bacteria requiring oxygen to function (aerobic bacteria) and fungi manage the chemical process by converting the inputs into heat, carbon dioxide and ammonium. The ammonium (NH_4) is the form of nitrogen used by plants. When available ammonium is not used by plants it is further converted by bacteria into nitrates (NO_3) through the process of nitrification.

Compost is rich in nutrients. It is used in gardens, landscaping, horticulture, and agriculture. The compost itself is beneficial for the land in many ways, including as a soil conditioner, a fertilizer, addition of vital humus or humic acids, and as a natural pesticide for soil. In ecosystems, compost is useful for erosion control, land and stream reclamation, wetland construction, and as landfill cover. Organic ingredients intended for composting can alternatively be used to generate biogas through anaerobic digestion.

Terminology

Composting of waste is an aerobic (in the presence of air) method of decomposing solid wastes. The process involves decomposition of organic waste into humus known as compost which is a good fertiliser for plants. However,the term "composting" is used worldwide with differing meanings. Some composting textbooks narrowly define composting as being an aerobic form of decomposition, primarily by microbes. An alternative term to composting is "aerobic digestion", which in turn is also referred to as "wet composting".

For many people, composting is used to refer to several different types of biological process. In North America, "anaerobic composting" is still a common term for what much of the rest of the world and in technical publications people call "anaerobic digestion". The microbes used and the processes involved are quite different between composting and anaerobic digestion.

Ingredients

Home compost barrel in the Escuela Barreales, Santa Cruz, Chile

Carbon, Nitrogen, Oxygen, Water

Materials in a compost pile

Food scraps compost heap

Composting organisms require four equally important ingredients to work effectively:

- Carbon — for energy; the microbial oxidation of carbon produces the heat, if included at suggested levels.

  - High carbon materials tend to be brown and dry.

- Nitrogen — to grow and reproduce more organisms to oxidize the carbon.

  - High nitrogen materials tend to be green (or colorful, such as fruits and vegetables) and wet.

- Oxygen — for oxidizing the carbon, the decomposition process.

- Water — in the right amounts to maintain activity without causing anaerobic conditions.

Certain ratios of these materials will provide beneficial bacteria with the nutrients to work at a rate that will heat up the pile. In that process much water will be released as vapor ("steam"), and the oxygen will be quickly depleted, explaining the need to actively manage the pile. The hotter the pile gets, the more often added air and water is necessary; the air/water balance is critical to maintaining high temperatures (135°-160° Fahrenheit / 50° - 70° Celsius) until the materials are broken down. At the same time, too much air or water also slows the process, as does too much carbon (or too little nitrogen). Hot container composting focuses on retaining the heat to increase decomposition rate and produce compost quicker.

The most efficient composting occurs with an optimal carbon:nitrogen ratio of about 10:1 to 20:1. Rapid composting is favored by having a C/N ratio of ~30 or less. Theoretical analysis is confirmed by field tests that above 30 the substrate is nitrogen starved, below 15 it is likely to outgas a portion of nitrogen as ammonia. If nitrogen needs to be increased, it has been suggested to add 0.15 pounds of *actual* nitrogen per three bushels (3.75 cubic feet) of lower nitrogen material. [For those not familiar with these types of units: 0.64g/L or 640 grams of actual nitrogen per cubic meter.] Two to 3 pounds of organic nitrogen supplement (blood meal, manure, bone meal, alfalfa meal) per 100 pounds of low nitrogen materials (for example, straw or sawdust), supplies generally ample nitrogen and trace minerals in high carbon mixes.

Nearly all plant and animal materials have both carbon and nitrogen, but amounts vary widely, with characteristics noted above (dry/wet, brown/green). Fresh grass clippings have an average ratio of about 15:1 and dry autumn leaves about 50:1 depending on species. Mixing equal parts by volume approximates the ideal C:N range. Few individual situations will provide the ideal mix of materials at any point. Observation of amounts, and consideration of different materials as a pile is built over time, can quickly achieve a workable technique for the individual situation.

Animal Manure and Bedding

On many farms, the basic composting ingredients are animal manure generated on the farm and bedding. Straw and sawdust are common bedding materials. Non-traditional bedding materials are also used, including newspaper and chopped cardboard. The amount of manure composted on a livestock farm is often determined by cleaning schedules, land availability, and weather conditions. Each type of manure has its own physical, chemical, and biological characteristics. Cattle and horse manures, when mixed with bedding, possess good qualities for composting. Swine manure, which is very wet and usually not mixed with bedding material, must be mixed with straw or similar raw materials. Poultry manure also must be blended with carbonaceous materials - those low in nitrogen preferred, such as sawdust or straw.

Microorganisms

With the proper mixture of water, oxygen, carbon, and nitrogen, micro-organisms are allowed to break down organic matter to produce compost. The composting process is dependent on micro-organisms to break down organic matter into compost. There are many types of microorganisms found in active compost of which the most common are:

- Bacteria- The most numerous of all the microorganisms found in compost. Depending on the phase of composting, mesophilic or thermophilic bacteria may predominate.

- Actinobacteria- Necessary for breaking down paper products such as newspaper, bark, etc.

- Fungi- Molds and yeast help break down materials that bacteria cannot, especially lignin in woody material.

- Protozoa- Help consume bacteria, fungi and micro organic particulates.

- Rotifers- Rotifers help control populations of bacteria and small protozoans.

In addition, earthworms not only ingest partly composted material, but also continually re-create aeration and drainage tunnels as they move through the compost.

A lack of a healthy micro-organism community is the main reason why composting processes are slow in landfills with environmental factors such as lack of oxygen, nutrients or water being the cause of the depleted biological community.

Phases of Composting

Under ideal conditions, composting proceeds through three major phases:

- An initial, mesophilic phase, in which the decomposition is carried out under moderate temperatures by mesophilic microorganisms.

- As the temperature rises, a second, thermophilic phase starts, in which the decomposition is carried out by various thermophilic bacteria under high temperatures.

- As the supply of high-energy compounds dwindles, the temperature starts to decrease, and the mesophiles once again predominate in the maturation phase.

Human Waste

Human waste (excreta) can also be added as an input to the composting process, like it is done in composting toilets, as human waste is a nitrogen-rich organic material.

People excrete far more water-soluble plant nutrients (nitrogen, phosphorus, potassium) in urine than in feces. Human urine can be used directly as fertilizer or it can be put onto compost. Adding a healthy person's urine to compost usually will increase temperatures and therefore increase its ability to destroy pathogens and unwanted seeds. Urine from a person with no obvious symptoms of infection is much more sanitary than fresh feces. Unlike feces, urine does not attract disease-spreading flies (such as house flies or blow flies), and it does not contain the most hardy of pathogens, such as parasitic worm eggs. Urine usually does not stink for long, particularly when it is fresh, diluted, or put on sorbents.

Urine is primarily composed of water and urea. Although metabolites of urea are nitrogen fertilizers, it is easy to over-fertilize with urine, or to utilize urine containing pharmaceutical (or other) content, creating too much ammonia for plants to absorb, acidic conditions, or other phytotoxicity.

Humanure

"Humanure" is a portmanteau of *human* and *manure*, designating human excrement (feces and urine) that is recycled via composting for agricultural or other purposes. The term was first used in a 1994 book by Joseph Jenkins that advocates the use of this organic soil amendment. The term humanure is used by compost enthusiasts in the US but not generally elsewhere. Because the term "humanure" has no authoritative definition it is subject to various uses; news reporters occasionally fail to correctly distinguish between humanure and sewage sludge or "biosolids".

Uses

Compost is generally recommended as an additive to soil, or other matrices such as coir and peat, as a tilth improver, supplying humus and nutrients. It provides a rich *growing medium*, or a porous, absorbent material that holds moisture and soluble minerals, providing the support and nutrients in which plants can flourish, although it is rarely used alone, being primarily mixed with soil, sand, grit, bark chips, vermiculite, perlite, or clay granules to produce loam. Compost can be tilled directly into the soil or growing medium to boost the level of organic matter and the overall fertility of the soil. Compost that is ready to be used as an additive is dark brown or even black with an earthy smell.

Generally, direct seeding into a compost is not recommended due to the speed with which it may dry and the possible presence of phytotoxins that may inhibit germination, and the possible tie up of nitrogen by incompletely decomposed lignin. It is very common to see blends of 20–30% compost used for transplanting seedlings at cotyledon stage or later.

Composting can destroy pathogens or unwanted seeds. Unwanted living plants (or weeds) can be discouraged by covering with mulch/compost. The "microbial pesticides" in compost may include thermophiles and mesophiles, however certain composting detritivores such as black soldier fly larvae and redworms, also reduce many pathogens. Thermophilic (high-temperature) composting is well known to destroy many seeds and nearly all types of pathogens (exceptions may include prions). The sanitizing qualities of (thermophilic) composting are desirable where there is a high likelihood of pathogens, such as with manure.

Composting Technologies

A homemade compost tumbler

A modern compost bin constructed from plastics

Overview

In addition to the traditional compost pile, various approaches have been developed to handle different composting processes, ingredients, locations, and applications for the composted product.

There is a large number of different composting systems on the market, for example:

- At the household level: Composting toilet, container composting, vermicomposting

- At the industrial composting (large scale): Aerated Static Pile Composting, vermicomposting, windrow composting etc.

Examples

Vermicomposting

Rotary screen harvested worm castings

Vermicompost is the product or process of composting through the utilization of various species of worms, usually red wigglers, white worms, and earthworms, to create a heterogeneous mixture of decomposing vegetable or food waste (excluding meat, dairy, fats, or oils), bedding materials, and vermicast. Vermicast, also known as worm castings, worm humus or worm manure, is the end-product of the breakdown of organic matter by species of earthworm. Vermicomposting is widely used in North America for on-site institutional processing of food waste, such as in hospitals and shopping malls. This type of composting is sometimes suggested as a feasible indoor home composting method. Vermicomposting has gained popularity in both these industrial and domestic settings because, as compared with conventional composting, it provides a way to compost organic materials more quickly (as defined by a higher rate of carbon-to-nitrogen ratio increase) and to attain products that have lower salinity levels that are therefore more beneficial to plant mediums.

Food waste - after three years

The earthworm species (or composting worms) most often used are red wigglers (*Eisenia fetida* or *Eisenia andrei*), though European nightcrawlers (*Eisenia hortensis* or *Dendrobaena veneta*) could also be used. Red wigglers are recommended by most vermiculture experts, as they have some of the best appetites and breed very quickly. Users refer to European nightcrawlers by a variety of other names, including *dendrobaenas, dendras,* Dutch Nightcrawlers, and Belgian nightcrawlers.

Containing water-soluble nutrients, vermicompost is a nutrient-rich organic fertilizer and soil conditioner in a form that is relatively easy for plants to absorb. Worm castings are sometimes used as an organic fertilizer. Because the earthworms grind and uniformly mix minerals in simple forms, plants need only minimal effort to obtain them. The worms' digestive systems also add beneficial microbes to help create a "living" soil environment for plants.

Vermicompost tea in conjunction with 10% castings has been shown to cause up to a 1.7 times growth in plant mass over plants grown without.

Researchers from the Pondicherry University discovered that worm composts can also be used to clean up heavy metals. The researchers found substantial reductions in heavy metals when the worms were released into the garbage and they are effective at removing lead, zinc, cadmium, copper and manganese.

Hügelkultur (Raised Garden Beds or Mounds)

An almost completed Hügelkultur bed; the bed does not have dirt on it yet.

The practice of making raised garden beds or mounds filled with rotting wood is also called "Hügelkultur" in German. It is in effect creating a Nurse log that is covered with dirt.

Benefits of hügelkultur garden beds include water retention and warming of soil. Buried wood becomes like a sponge as it decomposes, able to capture water and store it for later use by crops planted on top of the hügelkultur bed.

The buried decomposing wood will also give off heat, as all compost does, for several years. These effects have been used by Sepp Holzer to enable fruit trees to survive at otherwise inhospitable temperatures and altitudes.

Black Soldier Fly Larvae Composting

Black Soldier Fly (*Hermetia illucens*) larvae have been shown to be able to rapidly consume large amounts of organic waste when kept at 31.8 °C, the optimum temperature for reproduction. Enthusiasts have experimented with a large number of different waste products and some even sell starter kits to the public.

Cockroach Composting

Cockroach composting is another insect-mediated composting method. In this case the adults of any number of cockroach species (such as the Turkestan cockroach or *Blaptica dubia*) are used to quickly convert manure or kitchen waste to nutrient dense compost. Depending on species used and environmental conditions, excess composting insects can be used as an excellent animal feed for farm animals and pets.

Bokashi

Inside a recently started bokashi bin. The aerated base is just visible through the food scraps and bokashi bran.

Bokashi is a method that uses a mix of microorganisms to cover food waste or wilted plants to decrease smell. Bokashi (ぼかし) is Japanese for "shading off" or "gradation." It derives from the practice of Japanese farmers centuries ago of covering food waste with rich, local soil that contained the microorganisms that would ferment the waste. After a few weeks, they would bury the waste.

Most practitioners obtain the microorganisms from the product Effective Microorganisms (EM1), first sold in the 1980s. EM1 is mixed with a carbon base (e.g. sawdust or bran) that it sticks to and a sugar for food (e.g. molasses). The mixture is layered with waste in a sealed container and after a few weeks, removed and buried.

Newspaper fermented in a lactobacillus culture can be substituted for bokashi bran for a successful bokashi bucket.

Compost Tea

Compost teas are defined as water extracts brewed from composted materials and can be derived from aerobic or anaerobic processes. Compost teas are generally produced from adding one volume of compost to 4-10 volumes of water, but there has also been debate about the benefits of aerating the mixture. Field studies have shown the benefits of adding compost teas to crops due to the adding of organic matter, increased nutrient availability and increased microbial activity. They have also been shown to have an effect on plant pathogens.

Composting Toilets

A composting toilet does not require water or electricity, and when properly managed does not smell. A composting toilet collects human excreta which is then added to a compost heap together with sawdust and straw or other carbon rich materials, where pathogens are destroyed to some extent. The amount of pathogen destruction depends on the temperature (mesophilic or thermophilic conditions) and composting time. A composting toilet tries to process the excreta in situ although this is often coupled with a secondary external composting step. The resulting compost product has been given various names, such as humanure and EcoHumus.

A composting toilet can aid in the conservation of fresh water by avoiding the usage of potable water required by the typical flush toilet. It further prevents the pollution of ground water by controlling the fecal matter decomposition before entering the system. When properly managed, there should be no ground contamination from leachate.

Compost and Land-Filling

As concern about landfill space increases, worldwide interest in recycling by means of composting is growing, since composting is a process for converting decomposable organic materials into useful stable products. Composting is one of the only ways to revitalize soil vitality due to phosphorus depletion in soil. Industrial scale composting in the form of in-vessel composting, aerated static pile composting, and anaerobic digestion takes place in most Western countries now, and in many areas is mandated by law. There are process and product guidelines in Europe that date to the early 1980s (Germany, the Netherlands, Switzerland) and only more recently in the UK and the US. In both these countries, private trade associations within the industry have established loose standards, some say as a stop-gap measure to discourage independent government agencies from establishing tougher consumer-friendly standards. The USA is the only Western country that does not distinguish sludge-source compost from green-composts, and by default in the USA 50% of states expect composts to comply in some manner with the federal EPA 503 rule promulgated in 1984 for sludge products. Compost is regulated in Canada and Australia as well.

Industrial Systems

A large compost pile that is steaming with the heat generated by thermophilic microorganisms.

Industrial composting systems are increasingly being installed as a waste management alternative to landfills, along with other advanced waste processing systems. Mechanical sorting of mixed waste streams combined with anaerobic digestion or in-vessel composting is called mechanical biological treatment, and is increasingly being used in developed countries due to regulations controlling the amount of organic matter allowed in landfills. Treating biodegradable waste before it enters a landfill reduces global warming from fugitive methane; untreated waste breaks down anaerobically in a landfill, producing landfill gas that contains methane, a potent greenhouse gas.

Vermicomposting, also known as vermiculture, is used for medium-scale on-site institutional composting, such as for food waste from universities and shopping malls. It is selected either as a more environmentally friendly choice than conventional methods of disposal, or to reduce the cost of commercial waste removal.

Large-scale composting systems are used by many urban areas around the world. Co-composting is a technique that combines solid waste with de-watered biosolids, although difficulties controlling inert and plastics contamination from municipal solid waste makes this approach less attractive. The world's largest MSW co-composter is the Edmonton Composting Facility in Edmonton, Alberta, Canada, which turns 220,000 tonnes of residential solid waste and 22,500 dry tonnes of biosolids per year into 80,000 tonnes of compost. The facility is 38,690 m² (416,500 sq.ft.) in area, equivalent to 4½ Canadian football fields, and the operating structure is the largest stainless steel building in North America, the size of 14 NHL rinks. In 2006, Qatar awarded Keppel Seghers Singapore, a subsidiary of Keppel Corporation, a contract to begin construction on a 275,000 tonne/year anaerobic digestion and composting plant licensed by Kompogas (de) Switzerland. This plant, with 15 independent anaerobic digesters, will be the world's largest composting facility once fully operational in early 2011 and forms part of Qatar's Domestic Solid Waste Management Centre, the largest integrated waste management complex in the Middle East.

Another large MSW composter is the Lahore Composting Facility in Lahore, Pakistan, which has a capacity to convert 1,000 tonnes of municipal solid waste per day into compost. It also has a capacity to convert substantial portion of the intake into refuse-derived fuel (RDF) materials for further combustion use in several energy consuming industries across Pakistan, for example in cement manufacturing companies where it is used to heat cement kilns. This project has also been approved by the Executive Board of the United Nations Framework Convention on Climate Change for reducing methane emissions, and has been registered with a capacity of reducing 108,686 tonnes CO_2 equivalent per annum.

Related Technologies

Anaerobic digestion is process for converting organic waste into (biogas). The residual material, sometimes in combination with sewage sludge can be followed by an aerobic composting process before selling or giving away the compost.

History

Compost Basket

Composting as a recognized practice dates to at least the early Roman Empire since Pliny the Elder (AD 23-79). Traditionally, composting involved piling organic materials until the next planting season, at which time the materials would have decayed enough to be ready for use in the soil. The advantage of this method is that little working time or effort is required from the composter and it fits in naturally with agricultural practices in temperate climates. Disadvantages (from the modern perspective) are that space is used for a whole year, some nutrients might be leached due to exposure to rainfall, and disease-producing organisms and insects may not be adequately controlled.

Composting was somewhat modernized beginning in the 1920s in Europe as a tool for organic farming. The first industrial station for the transformation of urban organic materials into compost was set up in Wels, Austria in the year 1921. Early frequent citations for propounding composting within farming are for the German-speaking world Rudolf Steiner, founder of a farming method called biodynamics, and Annie Francé-Harrar, who was appointed on behalf of the government in Mexico and supported the country 1950–1958 to set up a large humus organization in the fight against erosion and soil degradation.

In the English-speaking world it was Sir Albert Howard who worked extensively in India on sustainable practices and Lady Eve Balfour who was a huge proponent of composting. Composting was imported to America by various followers of these early European movements by the likes of J.I. Rodale (founder of Rodale Organic Gardening), E.E. Pfeiffer (who developed scientific practices in biodynamic farming), Paul Keene (founder of Walnut Acres in Pennsylvania), and Scott and Helen Nearing (who inspired the back-to-the-land movement of the 1960s). Coincidentally, some of the above met briefly in India - all were quite influential in the U.S. from the 1960s into the 1980s.

There are many modern proponents of rapid composting that attempt to correct some of the perceived problems associated with traditional, slow composting. Many advocate that compost can be made in 2 to 3 weeks. Many such short processes involve a few changes to traditional methods, including smaller, more homogenized pieces in the compost, controlling carbon-to-nitrogen ratio (C:N) at 30 to 1 or less, and monitoring the moisture level more carefully. However, none of these parameters differ significantly from the early writings of Howard and Balfour, suggesting that in fact modern composting has not made significant advances over the traditional methods that take a few months to work. For this reason and others, many modern scientists who deal with carbon transformations are sceptical that there is a "super-charged" way to get nature to make compost rapidly.

In fact, both sides are right to some extent. The bacterial activity in rapid high heat methods breaks down the material to the extent that pathogens and seeds are destroyed, and the original feedstock is unrecognizable. At this stage, the compost can be used to prepare fields or other planting areas. However, most professionals recommend that the compost be given time to cure before using in a nursery for starting seeds or growing young plants. The curing time allows fungi to continue the decomposition process and eliminating phytotoxic substances.

Many countries such as Wales and some individual cities such as Seattle and San Francisco require food and yard waste to be sorted for composting.

Kew Gardens in London has one of the biggest non-commercial compost heaps in Europe.

Drip Irrigation

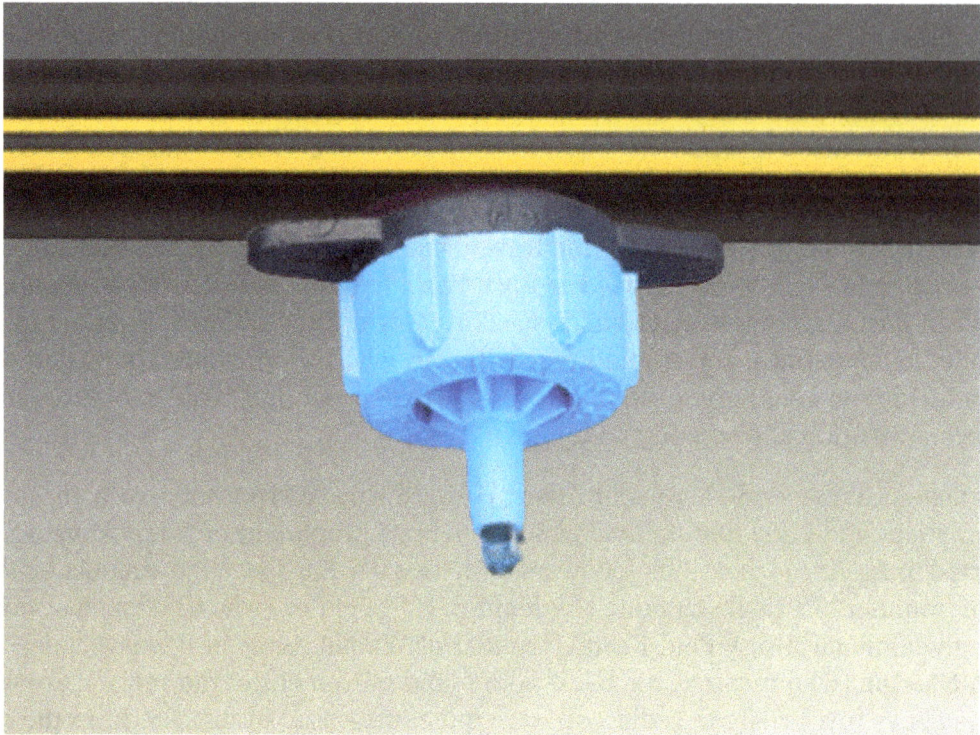

An Emitter or dripper in action

Drip irrigation is a form of irrigation that saves water and fertilizer by allowing water to drip slowly to the roots of many different plants, either onto the soil surface or directly onto the root zone, through a network of valves, pipes, tubing, and emitters. It is done through narrow tubes that deliver water directly to the base of the plant. It is chosen instead of surface irrigation for various reasons, often including concern about minimizing evaporation.

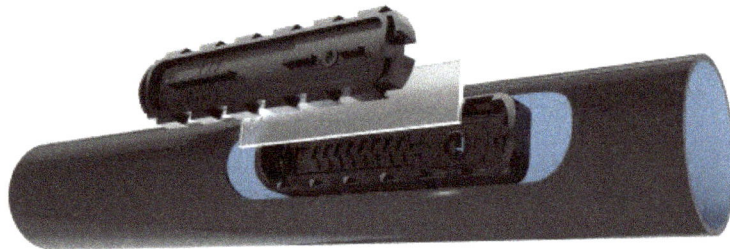

Open pressure compensated dripper

History

Primitive drip irrigation has been used since ancient times. Fan Sheng-Chih Shu, written in China during the first century BCE, describes the use of buried, unglazed clay pots filled with water as a means of irrigation. Modern drip irrigation began its development in Germany in 1860 when researchers began experimenting with subsurface irrigation using clay pipe to create combination irrigation and drainage systems. Research was later expanded in the 1920s to include the application of perforated pipe systems. The usage of plastic to hold and distribute water in drip irrigation was later developed in Australia by Hannis Thill.

Drip irrigation in Mexico vineyard, 2000

Usage of a plastic emitter in drip irrigation was developed in Israel by Simcha Blass and his son Yeshayahu. Instead of releasing water through tiny holes easily blocked by tiny particles, water was released through larger and longer passageways by using velocity to slow water inside a plastic emitter. The first experimental system of this type was established in 1959 by Blass who partnered later (1964) with Kibbutz Hatzerim to create an irrigation company called Netafim. Together they developed and patented the first practical surface drip irrigation emitter.

In the United States, the first drip tape, called *Dew Hose*, was developed by Richard Chapin of Chapin Watermatics in the early 1960s.

Modern drip irrigation has arguably become the world's most valued innovation in agriculture since the invention of the impact sprinkler in the 1930s, which offered the first practical alternative to surface irrigation. Drip irrigation may also use devices called micro-spray heads, which spray water in a small area, instead of dripping emitters. These are generally used on tree and vine crops with wider root zones. Subsurface drip irrigation (SDI) uses permanently or temporarily buried dripperline or drip tape located at or below the plant roots. It is becoming popular for row crop irrigation, especially in areas where water supplies are limited or recycled water is used for irrigation. Careful study of all the relevant factors like land topography, soil, water, crop and agro-climatic conditions are needed to determine the most suitable drip irrigation system and components to be used in a specific installation.

Components and Operation

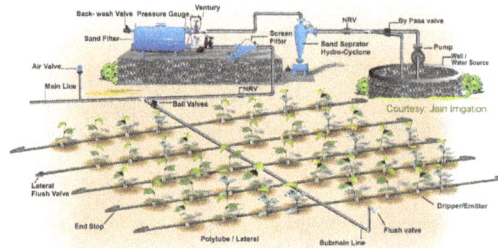

Drip irrigation system layout and its parts

Water distribution in subsurface drip irrigation

Nursery flowers watered with drip irrigation in Israel

Horticulture drip emitter in a pot

Components used in drip irrigation (listed in order from water source) include:

- Pump or pressurized water source

- Water filter(s) or filtration systems: sand separator, Fertigation systems (Venturi injector) and chemigation equipment (optional)

- Backwash controller (Backflow prevention device)

- Pressure Control Valve (pressure regulator)

- Main line (larger diameter pipe and pipe fittings)

- Hand-operated, electronic, or hydraulic control valves and safety valves

- Smaller diameter polytube (often referred to as "laterals")

- Poly fittings and accessories (to make connections)

- Emitting devices at plants (emitter or dripper, micro spray head, inline dripper or inline driptube)

In drip irrigation systems, pump and valves may be manually or automatically operated by a controller.

Most large drip irrigation systems employ some type of filter to prevent clogging of the small emitter flow path by small waterborne particles. New technologies are now being offered that minimize clogging. Some residential systems are installed without additional filters since potable water is already filtered at the water treatment plant. Virtually all drip irrigation equipment manufacturers recommend that filters be employed and generally will not honor warranties unless this is done. Last line filters just before the final delivery pipe are strongly recommended in addition to any other filtration system due to fine particle settlement and accidental insertion of particles in the intermediate lines.

Drip and subsurface drip irrigation is used almost exclusively when using recycled municipal waste water. Regulations typically do not permit spraying water through the air that has not been fully treated to potable water standards.

Because of the way the water is applied in a drip system, traditional surface applications of timed-release fertilizer are sometimes ineffective, so drip systems often mix liquid fertilizer with the irrigation water. This is called fertigation; fertigation and chemigation (application of pesticides and other chemicals to periodically clean out the system, such as chlorine or sulfuric acid) use chemical injectors such as diaphragm pumps, piston pumps, or aspirators. The chemicals may be added constantly whenever the system is irrigating or at intervals. Fertilizer savings of up to 95% are being reported from recent university field tests using drip fertigation and slow water delivery as compared to timed-release and irrigation by micro spray heads.

Properly designed, installed, and managed, drip irrigation may help achieve water conservation by reducing evaporation and deep drainage when compared to other types of irrigation such as flood or overhead sprinklers since water can be more precisely applied to the plant roots. In addition, drip can eliminate many diseases that are spread through water contact with the foliage. Finally, in regions where water supplies are severely limited, there may be no actual water savings, but rather simply an increase in production while using the same amount of water as before. In very arid regions or on sandy soils, the preferred method is to apply the irrigation water as slowly as possible.

Pulsed irrigation is sometimes used to decrease the amount of water delivered to the plant at any one time, thus reducing runoff or deep percolation. Pulsed systems are typically expensive and require extensive maintenance. Therefore, the latest efforts by emitter manufacturers are focused toward developing new technologies that deliver irrigation water at ultra-low flow rates, i.e. less than 1.0 liter per hour. Slow and even delivery further improves water use efficiency without incurring the expense and complexity of pulsed delivery equipment.

An emitting pipe is a type of drip irrigation tubing with emitters pre-installed at the factory with specific distance and flow per hour as per crop distance.

An emitter restricts water flow passage through it, thus creating head loss required (to the extent of atmospheric pressure) in order to emit water in the form of droplets. This head loss is achieved by friction / turbulence within the emitter.

Advantages and Disadvantages

Drip irrigation and spare drip irrigation tubes in banana farm at Chinawal, India

The advantages of drip irrigation are:

- Fertilizer and nutrient loss is minimized due to localized application and reduced leaching.
- Water application efficiency is high if managed correctly

- Field levelling is not necessary.

- Fields with irregular shapes are easily accommodated.

- Recycled non-potable water can be safely used.

Pot irrigation by On-line drippers

Pressure compensated integral dripper on soilless growing channels

- Moisture within the root zone can be maintained at field capacity.

- Soil type plays less important role in frequency of irrigation.

- Soil erosion is lessened.

- Weed growth is lessened.

- Water distribution is highly uniform, controlled by output of each nozzle.

- Labour cost is less than other irrigation methods.

- Variation in supply can be regulated by regulating the valves and drippers.

- Fertigation can easily be included with minimal waste of fertilizers.

- Foliage remains dry, reducing the risk of disease.

- Usually operated at lower pressure than other types of pressurised irrigation, reducing energy costs.

The disadvantages of drip irrigation are:

- Initial cost can be more than overhead systems.

- The sun can affect the tubes used for drip irrigation, shortening their usable life. (This article does not include a discussion of the effects of degrading plastic on the soil content and subsequent effect on food crops. With many types of plastic, when the sun degrades the plastic, causing it to become brittle, the estrogenic chemicals (that is, chemicals replicating female hormones) which would cause the plastic to retain flexibility have been released into the surrounding environment.)

- If the water is not properly filtered and the equipment not properly maintained, it can result in clogging.

- For subsurface drip the irrigator cannot see the water that is applied. This may lead to the farmer either applying too much water (low efficiency) or an insufficient amount of water, this is particularly common for those with less experience with drip irrigation.

- Drip irrigation might be unsatisfactory if herbicides or top dressed fertilizers need sprinkler irrigation for activation.

- Drip tape causes extra cleanup costs after harvest. Users need to plan for drip tape winding, disposal, recycling or reuse.

- Waste of water, time and harvest, if not installed properly. These systems require careful study of all the relevant factors like land topography, soil, water, crop and agro-climatic conditions, and suitability of drip irrigation system and its components.

- In lighter soils subsurface drip may be unable to wet the soil surface for germination. Requires careful consideration of the installation depth.

- most drip systems are designed for high efficiency, meaning little or no leaching fraction. Without sufficient leaching, salts applied with the irrigation water may build up in the root zone, usually at the edge of the wetting pattern. On the other hand, drip irrigation avoids the high capillary potential of traditional surface-applied irrigation, which can draw salt deposits up from deposits below.

- the PVC pipes often suffer from rodent damage, requiring replacement of the entire tube and increasing expenses.

- Drip irrigation systems cannot be used for damage control by night frosts (like in the case of sprinkler irrigation systems)

Uses

Irrigation dripper

Drip irrigation is used in farms, commercial greenhouses, and residential gardeners. Drip irrigation is adopted extensively in areas of acute water scarcity and especially for crops and trees such as coconuts, containerized landscape trees, grapes, bananas, pandey, eggplant, citrus, strawberries, sugarcane, cotton, maize, and potatoes.

Drip irrigation for garden available in drip kits are increasingly popular for the homeowner and consist of a timer, hose and emitter. Hoses that are 4 mm in diameter are used to irrigate flower pots.

Zero Waste Agriculture

Zero waste agriculture is a type of sustainable agriculture which optimizes use of the five natural kingdoms, i.e. plants, animals, bacteria, fungi and algae, to produce biodiverse-food, energy and nutrients in a synergistic integrated cycle of profit making processes where the waste of each process becomes the feedstock for another process.

Digester

The biogas digester is the heart of most zero waste agriculture (ZWA) systems. It is a 3000-year-old anaerobic digestion process.

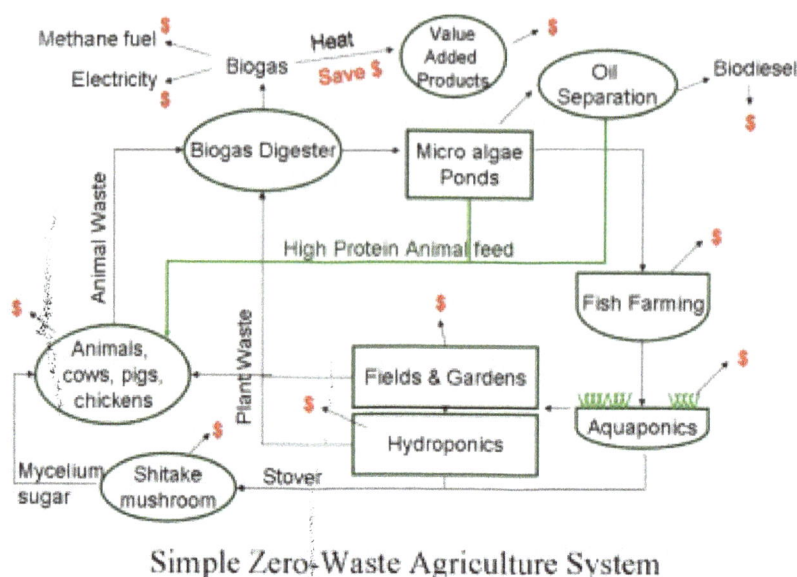

Simple Zero-Waste Agriculture System

History

The integration of shallow microaglal oxidisation ponds was demonstrated by Golueke & Oswald in the 1960s. The widespread global implementation of these systems can be largely credited to Prof George Chan from ZERI. Zero waste agriculture is now practiced in China (ecological farming), Columbia (integrated food & waste management systems) & Fiji (integrated farming systems), India (integrated biogas farming), South Africa (BEAT Coop & African Agroecological Biotechnology Initiative) and Mauritius. The Brazilian government has adopted integrated farming system as a major social technology for the uplifting of marginalized and subsistence farmers through coordination with TECPAR.

Zero waste agriculture combines mature ecological farming practices that delivers an integrated balance of job creation, poverty relief, food security, energy security, water conservation, climate change relief, land security & stewardship.

Practice

Zero waste agriculture is optimally practiced on small 1-5 ha sized family owned and managed farms and it complements traditional farming & animal husbandry as practiced in most third world communities. Zero Waste Agriculture also preserves local indigenous systems and existing agrarian cultural values and practices.

Zero waste agriculture presents a balance of economically, socially and ecologically benefits as it:

1. optimizes food production in an ecological sound manner

2. reduces water consumption through and recycling and reduced evaporation

3. provides energy security through the harvesting of biomethane (biogas) and the extraction

of biodiesel from micro-algae all of which from as a by-products of food production

4. provides climate change relief through the substantial reduction in greenhouse gas emissions from both traditional agriculture practices and fossil fuel usage

5. reduces the use of pesticides through biodiverse farming

Oil and Biodiesel from Algae

In sunny climates, a one hectare zero waste farm can produce over 1000 litres of oil in a year from the chlorella microalgae grown on biogas digester effluent in a 500m^2 shallow pond. The nutritive high protein waste from the oil extraction process can be used as an animal feed.

Multiple Cropping

In agriculture, multiple cropping is the practice of growing two or more crops in the same piece of land during a single growing season. It is a form of polyculture. It can take the form of double-cropping, in which a second crop is planted after the first has been harvested, or relay cropping, in which the second crop is started amidst the first crop before it has been harvested. A related practice, companion planting, is sometimes used in gardening and intensive cultivation of vegetables and fruits. One example of multi-cropping is tomatoes + onions + marigold; the marigolds repel some tomato pests.

Multiple cropping is found in many agricultural traditions. In the Garhwal Himalaya of India, a practice called baranaja involves sowing 12 or more crops on the same plot, including various types of beans, grains, and millets, and harvesting them at different times.

In the cultivation of rice, multiple cropping requires effective irrigation, especially in areas with a dry season. Rain that falls during the wet season permits the cultivation of rice during that period, but during the other half of the year, water cannot be channeled into the rice fields without an irrigation system. The Green Revolution in Asia led to the development of high-yield varieties of rice, which required a substantially shorter growing season of 100 days, as opposed to traditional varieties, which needed 150 to 180 days. Due to this, multiple cropping became more prevalent in Asian countries.

Afforestation

Afforestation is the establishment of a forest or stand of trees in an area where there was no previous tree cover. Reforestation is the reestablishment of forest cover, either naturally (by natural seeding, coppice, or root suckers) or artificially (by direct seeding or planting).

Forestation is the establishment of forest growth on areas that either had forest or lacked it. Reforestation and afforestation are categories of forestation. Many governments and non-governmental organizations directly engage in programs of *afforestation* to create forests, increase carbon capture and sequestration, and help to anthropogenically improve biodiversity. (In the UK, afforestation may mean converting the legal status of some land to "royal forest".) Special tools, e.g. tree planting bar, are used to make planting of trees easier and faster.

An afforestation project in Rand Wood, Lincolnshire, England

Biological Process

Gap dynamics is the pattern of plant growth that occurs following the creation of a forest gap, a local area of natural disturbance that results in an opening in the canopy of a forest. Gap dynamics are a typical characteristic of temperate and tropical forests, and have a wide variety of causes and effects on forest life.

In areas of Degraded Soil

In some places, forests need help to reestablish themselves because of environmental factors. For example, in arid zones, once forest cover is destroyed, the land may dry and become inhospitable to new tree growth. Other factors include overgrazing by livestock, especially animals such as goats, cows, and over-harvesting of forest resources. Together these may lead to desertification and the loss of topsoil; without soil, forests cannot grow until the long process of soil creation has been completed - if erosion allows this. In some tropical areas, forest cover removal may result in a duricrust or duripan that effectively seal off the soil to water penetration and root growth. In many areas, reforestation is impossible because people are using the land. In other areas, mechanical breaking up of duripans or duricrusts is necessary, careful and continued watering may be essential, and special protection, such as fencing, may be needed.

Countries and Regions

Afforested botanical garden in Hattori Ryokuchi Park, Japan.

Brazil

There is extensive and ongoing Amazon deforestation.)

China

China has deforested most of its historically wooded areas. China reached the point where timber yields declined far below historic levels, due to over-harvesting of trees beyond sustainable yield. Although it has set official goals for reforestation, these goals are set over an 80-year time horizon and have not been significantly met by 2008. China is trying to correct these problems by projects as the Green Wall of China, which aims to replant a great deal of forests and halt the expansion of the Gobi desert. A law promulgated in 1981 requires that every school student over the age of 11 plants at least one tree per year. As a result, China has the highest afforestation rate of any country or region in the world, with 47,000 square kilometers of afforestation in 2008. However, the forest area per capita is still far lower than the international average. There has also been considerable criticism regarding the effectiveness of planting so many trees especially in regions where they never grew prior. Studies reveal that the water table of those areas is becoming deeper indicating significant water loss.

India

India has witnessed a minor increase in the percentage of the land area under forest cover from 1950 to 2006. In 1950 around 40.48 million hectares was covered by forest. In 1980 it increased to 67.47 million hectares and in 2006 it was found to be 69 million hectares. 23% of India is covered by forest. The forests of India are grouped into 5 major categories and 16 types based on biophysical criteria. 38% of forest is categorised as subtropical dry deciduous and 30% as tropical moist deciduous plus other smaller groups. It is taken care that only local species are planted in an area. Trees bearing fruits are preferred wherever possible due to their function as a food source.

Afforestation in South India

Hong Kong

Since the founding of the crown colony in the 19th century, afforestation has taken place to prevent soil erosion in the catchment areas of the reservoirs that were built. During the Japanese occupation in the Second World War, the countryside was deforested as the remaining population required fuel to survive. Most of the trees were cut down and extensive reafforestation was carried out after the war. Trees that were planted are mostly non-native species, such as: Pinus massoniana, Acacia confusa (Formosan acacia), Lophostemon confertus and the Paper Bark Tree.

Burkina Faso

Desertification is increasing along the Sahel, the strip of land between Africa's fertile tropics and the Sahara Desert. After a crippling famine in the 1970s caused by overgrazing and deforestation, a local community approach has been pioneered by Yacouba Sawadogo, a peasant farmer. By replanting trees and crops together in holes filled with compost, whole villages have been able to move back to areas considered uninhabitable.

Iran

Iran is considered a low forest cover region of the world with present cover approximating seven percent of the land area. This is a value reduced by an estimated six million hectares of virgin forest, which includes oak, almond and pistachio. Due to soil substrates, it is difficult to achieve afforestation on a large scale compared to other temperate areas endowed with more fertile and less rocky and arid soil conditions. Consequently, most of the afforestation is conducted with non-native species, leading to habitat destruction for native flora and fauna, and resulting in an accelerated loss of biodiversity.

JNF trees in the Negev Desert. Man-made dunes (here a liman) help keep in rainwater, creating an oasis.

Israel

Tree-planting is an ancient Jewish tradition, mentioned in the Talmud as being more important than greeting the Messiah. With over 240 million planted trees, Israel is one of only two countries that entered the 21st century with a net gain in the number of trees, due to massive afforestation efforts. Israeli forests are the product of a major afforestation campaign by the Jewish National Fund (JNF).

Critics argue that many JNF lands inside the West Bank were illegally confiscated from Palestinian refugees, and that the JNF furthermore should not be involved with lands in the West Bank. Shaul Ephraim Cohen has claimed that trees have been planted to restrict Bedouin herding. Susan Nathan wrote that forests were planted on the site of abandoned Arab villages after the 1948 war.

Since 2009, the JNF has provided the Palestinian Authority with 3,000 tree seedlings for a forested area being developed on the edge of the new city of Rawabi, north of Ramallah.

North Africa

In North Africa, the Sahara Forest Project coupled with the Seawater greenhouse has been proposed. Some projects have also been launched in countries as Senegal to revert desertification.

As of 2010, African leaders are discussing the combining of national resources to increase effectiveness. In addition, other projects as the Keita Project in Niger have been launched in the past, and have been able to locally revert damage done by desertification.

Europe

Europe has deforested the majority of its historical forests. The European Union (EU) has paid farmers for afforestation since 1990, offering grants to turn farmland back into forest and payments for the management of forest. Between 1993 and 1997, EU afforestation policies made possible the re-forestation of over 5,000 square kilometres of land. A second program, running between 2000 and 2006, afforested more than 1000 square kilometres of land (precise statistics not yet available). A third such program began in 2007. Europe's forests are growing by 0.8 million ha a year thanks to these programmes.

In Poland, the National Program of Afforestation was introduced by the government after World War II, when area of forests shrank to 20% of country's territory. Consequently, forested areas of Poland grew year by year, and on December 31, 2006, forests covered 29% of the country. It is planned that by 2050, forests will cover 33% of Poland.

According to Food and Agriculture Organization statistics, Spain had the third fastest afforestation rate in Europe in the 1990-2005 period, after Iceland and Ireland. In those years, a total of 44,360 square kilometers were afforested, and the total forest cover rose from 13,5 to 17,9 million hectares. In 1990, forests covered 26.6% of the Spanish territory. As of 2007, that figure had risen to 36.6%. Spain today has the fifth largest forest area in the European Union.

In January 2013 the UK government set a target of 12% woodland cover in England by 2060, up from the then 10%. Government-backed initiatives such as the Woodland Carbon Code are intended to support this objective by encouraging corporations and landowners to create new woodland to offset their carbon emissions.

Alpine and Subalpine regions have undergone a lot of deforestation and then forestation in the last 300 years. Out of this has emerged much practical experience. One example is the Rotten group, which is a method to bring in stable age mixed tree communities.

Australia

In Adelaide, South Australia (a city of 1.3 million),Premier Mike Rann (2002 to 2011) launched an urban forest initiative in 2003 to plant 3 million native trees and shrubs by 2014 on 300 project sites across the metro area. The projects range from large habitat restoration projects to local biodiversity projects. Thousands of Adelaide citizens have participated in community planting days. Sites include parks, reserves, transport corridors, schools, water courses and coastline. Only trees native to the local area are planted to ensure genetic integrity. Premier Rann said the project aimed to beautify and cool the city and make it more liveable; improve air and water quality and reduce Adelaide's greenhouse gas emissions by 600,000 tonnes of CO_2 a year. He said it was also about creating and conserving habitat for wildlife and preventing species loss.

United States

The United States is roughly one-third covered in forest and woodland. Nevertheless, areas in the US were subject to significant tree planting. In the 1800s people moving westward encountered the Great Plains; land with fertile soil, a growing population and a demand for timber but with few trees to supply it. So tree planting was encouraged along homesteads. Arbor Day was founded in 1872 by Julius Sterling Morton in Nebraska City, Nebraska. By the 1930s the environmental disaster, the Dust Bowl signified a reason for significant tree cover. Public work's programs under the New Deal saw the planting of 18,000 miles of windbreaks stretching from North Dakota to Texas to fight soil erosion.

At their summit in Copenhagen in 2009, organised by the UK based The Climate Group, leaders of sub-national governments - States, Regions and Provinces - unanimously supported a recommendation by Premier Rann to plant 1 billion trees across their varied jurisdictions. The initiative was strongly supported by leaders present including Quebec Premier Jean Charest, California Governor Arnold Schwarzenegger and Scottish First Minister Alex Salmond. At a subsequent meeting in Rio de Janeiro in June 2012, The Climate Group announced that it had already received commitments by member governments to plant more than 500 million trees.

No-Till Farming

Young soybean plants thrive in and are protected by the residue of a wheat crop. This form of no till farming provides good protection for the soil from erosion and helps retain moisture for the new crop.

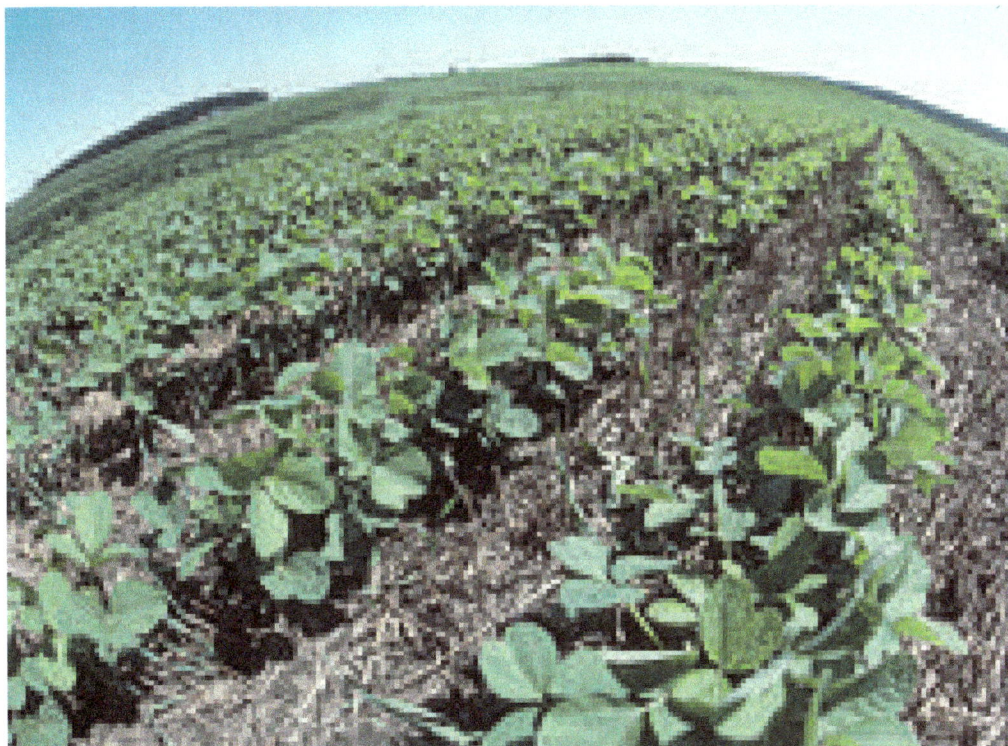

No-till farming (also called zero tillage or direct drilling) is a way of growing crops or pasture from year to year without disturbing the soil through tillage. No-till is an agricultural technique which increases the amount of water that infiltrates into the soil and increases organic matter retention and cycling of nutrients in the soil. In many agricultural regions it can reduce or eliminate soil erosion. It increases the amount and variety of life in and on the soil, including disease-causing organisms and disease suppression organisms. The most powerful benefit of no-tillage is improvement in soil biological fertility, making soils more resilient. Farm operations are made much more efficient, particularly improved time of sowing and better trafficability of farm operations.

Background

Tilling is the process of removing plants or plant debris, usually for the purposes of planting more desirable species. This tilling can result in a flat seed bed or one that has formed areas, such as rows or raised beds, to enhance the growth of desired plants. It is an ancient technique with clear evidence of its use since at least 3000 B.C.

The effects of tillage can include soil compaction; loss of organic matter; degradation of soil aggregates; death or disruption of soil microbes and other organisms including mycorrhiza, arthropods, and earthworms; and soil erosion where topsoil is washed or blown away.

Origin of No-Till for Modern Farms

The idea of modern no-till farming started in the 1940s with Edward H. Faulkner, author of *Plowman's Folly*, but it wasn't until the development of several chemicals after WWII that various researchers and farmers started to try out the idea. The first adopters of no-till include Klingman (North Carolina), Edward Faulkner, L.A. Porter (New Zealand), Harry and Lawrence Young (Herndon, Kentucky), the Instituto de Pesquisas Agropecuarias Meridional (1971 in Brazil) with Herbert Bartz.

Adoption Rate in the United States

No-till farming is widely used in the United States and the number of acres managed in this way continues to grow. This growth is supported by a decrease in costs related to tillage; no-till management results in fewer passes with equipment for approximately equal harvests, and the crop residue prevents evaporation of rainfall and increases water infiltration into the soil.

Issues

Profit, Economics, Yield

Studies have found that no-till farming can be more profitable if performed correctly.

Less tillage of the soil reduces labour, fuel, irrigation and machinery costs. No-till can increase yield because of higher water infiltration and storage capacity, and less erosion. Another benefit of no-till is that because of the higher water content, instead of leaving a field fallow it can make economic sense to plant another crop instead.

As sustainable agriculture becomes more popular, monetary grants and awards are becoming readily available to farmers who practice conservation tillage. Some large energy corporations which are among the greatest generators of fossil-fuel-related pollution may purchase carbon credits, which can encourage farmers to engage in conservation tillage. Under such schemes, the farmers' land is legally redefined as a carbon sink for the power generators' emissions. This helps the farmer in several ways, and it helps the energy companies meet regulatory demands for reduction of pollution, specifically carbon emissions.

No-till farming can increase organic (carbon based) matter in the soil, which is a form of carbon sequestration. However, there is debate over whether this increased sequestration detected in scientific studies of no-till agriculture is actually occurring, or is due to flawed testing methods or other factors. Regardless of this debate, there are still many other good reasons to use no-till, e.g. reduction in fossil fuel use, no erosion, soil quality.

Environmental

Carbon (Air and Soil) and Other Greenhouse Gases

No-till farming has carbon sequestration potential through storage of soil organic matter in the soil of crop fields. Whereas, when soil is tilled by machinery, the soil layers invert, air mixes in, and soil microbial activity dramatically increases over baseline levels. Tilling results in soil organic matter being broken down much more rapidly, and carbon is lost from the soil into the atmosphere. In addition to the effect on soil from tilling, emissions from the farm tractors increases carbon dioxide levels in the atmosphere.

Cropland soils are ideal for use as a carbon sink, since they have been depleted of carbon in most areas. It is estimated that 78 billion metric tonnes of carbon that was trapped in the soil has been released because of tillage. Conventional farming practices that rely on tillage have removed carbon from the soil ecosystem by removing crop residues such as left over corn stalks, and through the addition of chemical fertilizers which have the above-mentioned effects on soil microbes. By eliminating tillage, crop residues decompose where they lie, and by growing winter cover crops, carbon loss can be slowed and eventually reversed.

Nonetheless, a growing body of research is showing that no-till systems lose carbon stocks over time. Regarding a 2014 study of which he was principal investigator, University of Illinois soil scientist Ken Olson said this differing result occurs in part because tested soil samples need to include the full depth of rooting; 1–2 meters deep. He said, "That no-till subsurface layer is often losing more soil organic carbon stock over time than is gained in the surface layer." Also, there has not been a uniform definition of soil organic carbon sequestration among researchers. The study concludes, "Additional investments in SOC research is needed to better understand the agricultural management practices that are most likely to sequester SOC or at least retain more net SOC stocks."

In addition to keeping carbon in the soil, no-till farming reduces nitrous oxide (N_2O) emissions by 40-70%, depending on rotation. Nitrous oxide is a potent greenhouse gas, 300 times stronger than CO2, and stays in the atmosphere for 120 years. Fertilizing farmlands with (excessive) nitrogen increases the release of nitrous oxide.

Soil and water

No-till farming improves soil quality (soil function), carbon, organic matter, aggregates, protecting the soil from erosion, evaporation of water, and structural breakdown. A reduction in tillage passes helps prevent the compaction of soil.

Recently, researchers at the Agricultural Research Service of the United States Department of Agriculture found that no-till farming makes soil much more stable than plowed soil. Their conclusions draw from over 19 years of collaborated tillage studies. No-till stores more carbon in the soil and carbon in the form of organic matter is a key factor in holding soil particles together. The first inch of no-till soil is two to seven times less vulnerable than that of plowed soil. The practice of no-till farming is especially beneficial to Great Plains farmers because of its resistance to erosion.

Crop residues left intact help both natural precipitation and irrigation water infiltrate the soil where it can be used. The crop residue left on the soil surface also limits evaporation, conserving water for plant growth. Soil compaction and no tillage-pan, soil absorbs more water and plants are able to grow their roots deeper into the soil and suck up more water.

Tilling a field reduces the amount of water, via evaporation, around 1/3 to 3/4 inches (0.85 to 1.9 cm) per pass. By no-tilling, this water stays in the soil, available to the plants.

Soil Biota, Wildlife, Etc.

In no-till farming the soil is left intact and crop residue is left on the field. Therefore, soil layers, and in turn soil biota, are conserved in their natural state. No-tilled fields often have more beneficial insects and annelids, a higher microbial content, and a greater amount of soil organic material. Since there is no ploughing there is less airborne dust.

No-till farming increases the amount and variety of wildlife. This is the result of improved cover, reduced traffic and the reduced chance of destroying ground nesting birds and animals (plowing destroys all of them).

Albedo

Tillage lowers the albedo of croplands. The potential for global cooling as a result of increased Albedo in no till croplands is similar in magnitude to the biogeochemical (carbon sequestration) potential.

Historical Artifacts

Tilling regularly damages ancient structures under the soil such as long barrows. In the UK, half of the long barrows in Gloucestershire and almost all the burial mounds in Essex have been damaged. According to English Heritage modern tillage techniques have done as much damage in the last six decades as traditional tilling did in the previous six centuries. By using no-till methods these structures can be preserved and can be properly investigated instead of being destroyed.

Cost

Equipment

No-till farming requires specialized seeding equipment designed to plant seeds into undisturbed crop residues and soil. If the farmer has equipment designed for tillage farming, purchasing new equipment (seed drills for example) would be expensive and while the cost could be offset by selling off plows, etc. doing so is not usually done until the farmer decides to switch completely over (after trying it out for a few years). This would result in more money being invested into equipment in the short term (until old equipment is sold off).

Drainage

If a soil has poor drainage, it may need drainage tiles or other devices in order to help with the removal of excess water under no-till. Farmers should remember that water infiltration will improve after several years of a field being in no-till farming, so they may want to wait until 5–8 years have passed to see if the problems persists before deciding to invest in such an expensive system.

Gullies

Gullies can be a problem in the long-term. While much less soil is displaced by using no-till farming, any drainage gulleys that do form will get deeper each year since they aren't being smoothed out by plowing. This may necessitate either sod drainways, waterways, permanent drainways, cover crops, etc. Gully formation can be avoided entirely with proper water management practices, including the creation of swales on contour.

Increased Chemical Use

One of the purposes of tilling is to remove weeds. No-till farming does change weed composition drastically. Faster growing weeds may no longer be a problem in the face of increased competition, but shrubs and trees may begin to grow eventually.

Some farmers attack this problem with a "burn-down" herbicide such as glyphosate in lieu of tillage for seedbed preparation and because of this, no-till is often associated with increased chemical use in comparison to traditional tillage based methods of crop production. However, there are many agroecological alternatives to increased chemical use, such as winter cover crops and the mulch cover they provide, soil solarization or burning.

Management

No-till farming requires some different skills in order to do it successfully. As with any production system, if no-till isn't done correctly, yields can drop. A combination of technique, equipment, pesticides, crop rotation, fertilization, and irrigation have to be used for local conditions.

Cover crops

Cover crops are used occasionally in no-till farming to help control weeds and increase nutrients in the soil (by using legumes) or by using plants with long roots to pull mobile nutrients back up to the surface from lower layers of the soil. Farmers experimenting with organic no-till use cover crops instead of tillage for controlling weeds, and are developing various methods to kill the cover crops (rollers, crimper, choppers, etc.) so that the newly planted crops can get enough light, water, nutrients, etc.

Disease, Pathogens, Insects and the Use of Crop Rotations

With no-till farming, residue from the previous years crops lie on the surface of the field, cooling it and increasing the moisture. This can cause increased or decreased or variations of diseases that occur, but not necessarily at a higher or lower rate than conventional tillage. In order to help eliminate weed, pest and disease problems, crop rotations are used. By rotating the crops on a multi-year cycle, pests and diseases will decrease since the pests will no longer have a food supply to support their numbers.

Organic No-Till Technique: The Cardboard Method

Some farmers who prefer to pursue a chemical-free management practice often rely on the use of normal, non-dyed corrugated cardboard for use on seed-beds and vegetable areas. Used correctly, cardboard placed on a specific area can

1. keep important fungal hyphae and microorganisms in the soil intact

2. prevent recurring weeds from popping up

3. increase residual nitrogen and plant nutrients by top-composting plant residues and

4. create valuable topsoil that is well suited for next years seeds or transplants.

The plant residues (left over plant matter originating from cover crops, grass clippings, original plant life etc.) will rot while underneath the cardboard so long as it remains sufficiently moist. This rotting attracts worms and other beneficial microorganisms to the site of decomposition, and over a series of a few seasons (usually Spring-->Fall or Fall-->Spring) and up to a few years, will create a layer of rich topsoil. Plants can then be direct seeded into the soil come spring, or holes can be cut into the cardboard to allow for transplantation. Using this method in conjunction with other sustainable practices such as composting/vermicompost, cover crops and rotations are often considered beneficial to both land and those who take from it.

Water Issues

No-till farming dramatically reduces the amount of erosion in a field. While much less soil is displaced, any gullies that do form will get deeper each year instead of being smoothed out by regular plowing. This may necessitate either sod drainways, waterways, permanent drainways, cover crops, etc.

A problem that occurs in some fields is water saturation in soils. Switching to no-till farming will help the drainage issue because of the qualities of soil under continuous no-till include a higher water infiltration rate.

Equipment

It is very important to have planting equipment that can properly penetrate through the residue, into the soil and prepare a good seedbed. Switching to no-till reduces the maximum amount of power needed from farm tractors, which means that a farmer can farm under no-till with a smaller tractor than if he/she were tilling. Using a smaller, lighter tractor has the added benefit of reducing compaction.

Soil Temperature

Another problem that growers face is that in the spring the soil will take longer to warm and dry, which may delay planting to a less ideal future date. One reason why the soil is slower to warm is that the field absorbs less solar energy as the residue covering the soil is a much lighter color than the black soil which would be exposed in conventional tillage. This can be managed by using row cleaners on a planter. Since the soil can be cooler, harvest can occur a few days later than a conventionally tilled field. Note: A cooler soil is also a benefit because water doesn't evaporate as fast.

Residue

On some crops, like continuous no-till corn, the thickness of the residue on the surface of the field can become a problem without proper preparation and/or equipment.

Fertilizer

One of the most common yield reducers is nitrogen being immobilized in the crop residue, which can take a few months to several years to decompose, depending on the crop's C to N ratio and the local environment. Fertilizer needs to be applied at a higher rate during the transition period while the soil rebuilds its organic matter. The nutrients in the organic matter will be eventually released back into the soil, so this is only a concern during the transition time frame (4–5 years for Kansas, USA). An innovative solution to this problem is to integrate animal husbandry in various ways to aid in the decomposition cycle.

Misconceptions

Need to Fluff the Soil

Although no-till farming often causes a slight increase in soil bulk density, periodic tilling is not needed to "fluff" the soil back up. No-till farming mimics the natural conditions under which most soils formed more closely than any other method of farming, in that the soil is left undisturbed except to place seeds in a position to germinate.

Similar Terms

No-till farming is not equivalent to conservation tillage or strip tillage. Conservation tillage is a group of practices that reduce the amount of tillage needed. No-till and strip tillage are both forms of conservation tillage. No-till is the practice of never tilling a field. Tilling every other year is called rotational tillage.

Natural Farming

Masanobu Fukuoka, originator of the natural farming method

Natural farming is an ecological farming approach established by Masanobu Fukuoka (1913–2008), a Japanese farmer and philosopher, introduced in his 1975 book The One-Straw Revolution. Fukuoka described his way of farming as 自然農法 (*shizen nōhō*) in Japanese. It is also referred to as "the Fukuoka Method", "the natural way of farming" or "do-nothing farming". The title refers not to lack of effort, but to the avoidance of manufactured inputs and equipment. Natural farming is related to fertility farming, organic farming, sustainable agriculture, agroforestry, ecoagriculture and permaculture but should be distinguished from biodynamic agriculture.

The system exploits the complexity of living organisms that shape each particular ecosystem. Fukuoka saw farming both as a means of producing food and as an aesthetic or spiritual approach to life, the ultimate goal of which was, "the cultivation and perfection of human beings". He suggested that farmers could benefit from closely observing local conditions. Natural farming is a closed system, one that demands no human-supplied inputs and mimics nature.

Fukuoka's ideas challenged conventions that are core to modern agro-industries, instead promoting an approach that takes advantage of the local environment. Natural farming differs from conventional organic farming, which Fukuoka considered to be another modern technique that disturbs nature.

Fukuoka claimed that his approach prevents water pollution, biodiversity loss and soil erosion, while providing ample amounts of food.

Principles

Fukuoka distilled natural farming into five principles:

1. No tillage

2. No fertilizer

3. No pesticides or herbicides

4. No weeding

5. No pruning

Though many of his plant varieties and practices relate specifically to Japan and even to local conditions in subtropical western Shikoku, his philosophy and the governing principles of his farming systems have been applied from Africa to the temperate northern hemisphere. In India, natural farming is often referred to as "Rishi Kheti". In India natural farming or rishi kheti includes ancient vedic principles of farming including use of animal waste and herbs for controlling pests and promoting growth. The rishi 's or Indian sages use cow products like buttermilk, milk, curd and its waste urine for preparing growth promoters. The Rishi or Vedic farming is regarded as non -violent farming without any usage of chemical fertilizer and pesticides. They obtain high quality natural or organic produce having medicinal values.Today still small number of farmers in Madhya Pradesh, Punjab, Maharashtra and Andhra Pradesh, Tamil Nadu use this farming in India.

Principally, natural farming minimises human labour and adopts, as closely as practical, nature's production of foods such as rice, barley, daikon or citrus in biodiverse agricultural ecosystems. Without plowing, seeds germinate well on the surface if site conditions meet the needs of the seeds placed there. Fukuoka used the presence of spiders in his fields as a key performance indicator of sustainability.}

Fukuoka specifies that the ground remain covered by weeds, white clover, alfalfa, herbaceous legumes, and sometimes deliberately sown herbaceous plants. Ground cover is present along with grain, vegetable crops and orchards. Chickens run free in orchards and ducks and carp populate rice fields.

Periodically ground layer plants including weeds may be cut and left on the surface, returning their nutrients to the soil, while suppressing weed growth. This also facilitates the sowing of seeds in the same area because the dense ground layer hides the seeds from animals such as birds.

For summer rice and winter barley grain crops, ground cover enhances nitrogen fixation. Straw from the previous crop mulches the topsoil. Each grain crop is sown before the previous one is harvested by broadcasting the seed among the standing crop. Later, this method was reduced to a single direct seeding of clover, barley and rice over the standing heads of rice. The result is a denser crop of smaller, but highly productive and stronger plants.

Fukuoka's practice and philosophy emphasised small scale operation and challenged the need for mechanised farming techniques for high productivity, efficiency and economies of scale. While his family's farm was larger than the Japanese average, he used one field of grain crops as a small-scale example of his system.

Climax Ecosystems

Ladybirds consume aphids and are considered beneficial by natural farmers that apply biological control.

In ecology, climax ecosystems are mature ecosystems that have reached a high degree of stability, productivity and diversity. Natural farmers attempt to mimic those virtues, creating a comparable climax ecosystem, and employ advanced techniques such as intercropping, companion planting and integrated pest management.

No-Till

Natural farming recognizes soils as a fundamental natural asset. Ancient soils possess physical and chemical attributes that render them capable of generating and supporting life abundance. It can be argued that tilling actually degrades the delicate balance of a climax soil:

1. Tilling may destroy crucial physical characteristics of a soil such as *water suction*, its ability to send moisture upwards, even during dry spells. The effect is due to pressure differences between soil areas. Furthermore, tilling most certainly destroys soil horizons and hence disrupts the established flow of nutrients. A study suggests that reduced tillage preserves the crop residues on the top of the soil, allowing organic matter to be formed more easily and hence increasing the total organic carbon and nitrogen when compared to conventional tillage. The increases in organic carbon and nitrogen increase aerobic, facultative anaerobic and anaerobic bacteria populations.

2. Tilling over-pumps oxygen to local soil residents, such as bacteria and fungi. As a result, the chemistry of the soil changes. Biological decomposition accelerates and the microbiota mass increases at the expense of other organic matter, adversely affecting most plants, including trees and vegetables. For plants to thrive a certain quantity of organic matter (around 5%) must be present in the soil.

3. Tilling uproots all the plants in the area, turning their roots into food for bacteria and fungi. This damages their ability to aerate the soil. Living roots drill millions of tiny holes in the soil and thus provide oxygen. They also create room for beneficial insects and annelids (the phylum of worms). Some types of roots contribute directly to soil fertility by funding a mutualistic relationship with certain kinds of bacteria (most famously the rhizobium) that can fix nitrogen.

Fukuoka advocated avoiding any change in the natural landscape. This idea differs significantly from some recent permaculture practice that focuses on permaculture design, which may involve the change in landscape. For example, Sepp Holzer, an Austrian permaculture farmer, advocates the creation of terraces on slopes to control soil erosion. Fukuoka avoided the creation of terraces in his farm, even though terraces were common in China and Japan in his time. Instead, he prevented soil erosion by simply growing trees and shrubs on slopes.

Fertility Farming

In 1951, Newman Turner advocated the practice of "fertility farming", a system featuring the use of a cover crop, no tillage, no chemical fertilizers, no pesticides, no weeding and no composting. Although Turner was a commercial farmer and did not practice random seeding of seed balls, his "fertility farming" principles share similarities with Fukuoka's system of natural farming. Turner also advocated a "natural method" of animal husbandry.

Nature Farming

Japanese farmer and philosopher Mokichi Okada, conceived of a "no fertilizer" farming system in the 1930s that predated Fukuoka. Okada used the same Chinese characters, which are generally translated in English as "nature farming". Agriculture researcher Hu-lian Xu claims that "nature farming" is the correct literal translation of the Japanese term.

Good Agricultural Practice

Good agricultural practice (GAP) are specific methods which, when applied to agriculture, create food for consumers or further processing that is safe and wholesome. While there are numerous competing definitions of what methods constitute good agricultural practice there are several broadly accepted schemes that producers can adhere to.

Food and Agricultural Organization of the United Nations GAP

The Food and Agricultural Organization of the United Nations (FAO) uses good agricultural practice as a collection of principles to apply for on-farm production and post-production processes, resulting in safe and healthy food and non-food agricultural products, while taking into account economical, social and environmental sustainability.

GAPs may be applied to a wide range of farming systems and at different scales. They are applied through sustainable agricultural methods,

- Economically and efficiently produce sufficient (food security), safe (food safety) and nutritious food (food quality);*Research that works for developing countries and Australia. Retrieved 25 November 2007.*

GAPs require maintaining a common database on integrated production techniques for each of the major agro-ecological area, thus to collect, analyze and disseminate information of good practices in relevant geographical contexts.

Mamun Hasan Pondicherry University

United States Department of Agriculture GAP/GHP Program

The United States Department of Agriculture Agricultural Marketing Service currently operates an audit/certification program to verify that farms use good agricultural practice and/or good handling practice. This is a voluntary program typically utilized by growers and packers to satisfy contractual requirements with retail and food service buyers. The program was implemented in 2002 after the New Jersey Department of Agriculture petitioned USDA-AMS to implement an audit based program to verify conformance to the 1998 Food & Drug Administration publication entitled, "Guide to Minimize Microbial Food Safety Hazards for Fresh Fruits and Vegetables."

The program has been updated several times since 2002, and includes additional certification programs such as commodity specific audit programs for mushrooms, tomatoes, leafy greens, and cantaloupes. In 2009, USDA-AMS participated in the GAPs Harmonization Initiative which "harmonized" 14 of the major North American GAP audit standards, which in 2011 resulted in the release and implementation of the Produce GAPs Harmonized Food Safety Standard.

Smallholder Productivity

Demand for agricultural crops is expected to double as the world's population reaches 9.1 billion by 2050. Increasing the quantity and quality of food in response to growing demand will require increased agricultural productivity. Good agricultural practices, often in combination with effective input use, are one of the best ways to increase smallholder productivity. Many agribusinesss are building sustainable supply chains to increase production and improve quality.

Soil

- Reducing erosion by wind and water through hedging and ditching

- Application of fertilizers at appropriate moments and in adequate doses (i.e., when the plant needs the fertilizer), to avoid run-off.

- Maintaining or restoring soil organic content, by manure application, use of grazing, crop rotation

- Reduce soil compaction issues (by avoiding using heavy mechanical devices)

- Maintain soil structure, by limiting heavy tillage practices FAO : GAP : FAO GAP Principles : Soil

- *In situ* green manuring by growing pulse crops like cowpea, horse gram, sunn hemp etc.

Water

- Practice scheduled irrigation, with monitoring of plant needs, and soil water reserve status to avoid water loss by drainage

- Prevent soil salinization by limiting water input to needs, and recycling water whenever possible

- Avoid crops with high water requirements in a low availability region

- Avoid drainage and fertilizer run-off

- Maintain permanent soil covering, in particular in winter to avoid nitrogen run-off

- Manage carefully water table, by limiting heavy output of water

- Restore or maintain wetlands

- Provide good water points for livestock (FAO : GAP : FAO GAP Principles : Water)

- Harvest water *in situ* by digging catch pits, crescent bunds across slope

Animal Production, Health and Welfare

- Respect of animal well-being (freedom from hunger and thirst; freedom from discomfort; freedom from pain, injury or disease; freedom to express normal behavior; and freedom from fear and distress)

- Avoid nontherapeutic mutilations, surgical or invasive procedures, such as tail docking and debeaking;

- Avoid negative impacts on landscape, environment and life: contamination of land for grazing, food, water and air

- Check stocks and flows, maintain structure of systems

- Prevent chemical and medical residues from entering the food chain

- Minimize non-therapeutic use of antibiotics or hormones

- Avoid feeding animals with animal wastes or animal matter (reducing the risk of alien viral or transgenic genes, or prions such as mad cow disease),

- Minimize transport of live animals (by foot, rail or road) (reducing the risk of epidemics, e.g., foot and mouth disease)

- Prevent waste run-off (e.g. nitrate contamination of water tables from pigs), nutrient loss and greenhouse gas emissions (methane from cows)

- Prefer safety measures standards in manipulation of equipment

- Apply traceability processes on the whole production chain (breeding, feed, medical treatment...) for consumer security and feedback possibility in case of a food crisis (e.g., dioxin). FAO : GAP : FAO GAP Principles : Animal Health and Welfare

Healthcare and Public Health

- quality assurance of the horticultural or agricultural production of medicinal plant *Máthé, A.; I. Máthé. "Quality assurance of cultivated and gathered medicinal plants". Retrieved 23 May 2009., World Health Organization (2003). "WHO guidelines on good agricultural and collection practices (GACP) for medicinal plants" (PDF). Retrieved 23 May 2009.*

Integrated Farming

Integrated Farming (IF) is a whole farm management system which aims to deliver more sustainable agriculture. It is a dynamic approach which can be applied to any farming system around the world. It involves attention to detail and continuous improvement in all areas of a farming business through informed management processes. Integrated Farming combines the best of modern tools and technologies with traditional practices according to a given site and situation. In simple words, it means using many ways of cultivation in a small space or land.

Definition

The holistic approach: Integrated Farming looks at and relates to the whole farm

The International Organisation of Biological Control (IOBC) describes Integrated Farming as a farming system where high quality food, feed, fibre and renewable energy are produced by using resources such as soil, water, air and nature as well as regulating factors to farm sustainably and with as little polluting inputs as possible.

Particular emphasis is placed on a holistic management approach looking at the whole farm as cross-linked unit, on the fundamental role and function of agro-ecosystems, on nutrient cycles which are balanced and adapted to the demand of the crops, and on health and welfare of all livestock on the farm. Preserving and enhancing soil fertility, maintaining and improving a diverse environment and the adherence to ethical and social criteria are indispensable basic elements. Crop protection takes into account all biological, technical and chemical methods which then are balanced carefully and with the objective to protect the environment, to maintain profitability of the business and fulfil social requirements.

EISA European Initiative for Sustainable Development in Agriculture e. V. have an Integrated Farming Framework which provides additional explanations on key aspects of Integrated Farming. These include: Organisation & Planning, Human & Social Capital, Energy Efficiency, Water Use & Protection, Climate Change & Air Quality, Soil Management, Crop Nutrition, Crop Health & Protection, Animal Husbandry, Health & Welfare, Landscape & Nature Conservation and Waste Management Pollution Control.

LEAF (Linking Environment and Farming) in the UK promotes a comparable model and defines Integrated Farm Management (IFM) as whole farm business approach that delivers sustainable farming.

Classification

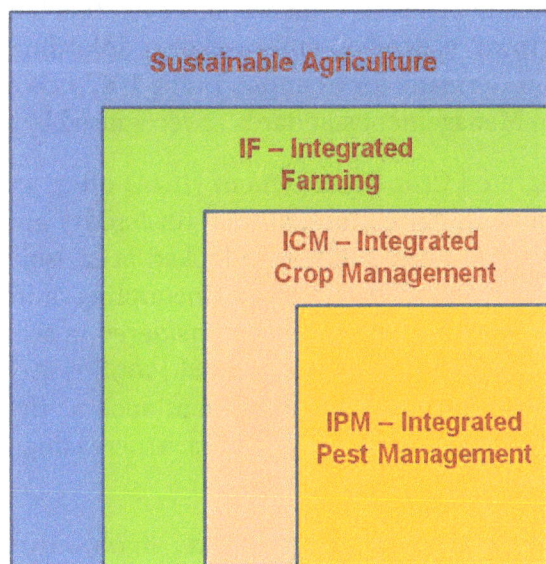

Integrated Farming in the context of sustainable agriculture

The Food and Agriculture Organization of the United Nations FAO promotes Integrated Pest Management (IPM) as the preferred approach to crop protection and regards it as a pillar of both sustainable intensification of crop production and pesticide risk reduction". IPM thus is one indispensable element of Integrated Crop Management which in turn is one essential part of the holistic Integrated Farming approach towards sustainable agriculture.

KELLER, 1985 (quoted in Lütke Entrup et al., 1998 1) highlights that Integrated Crop Management is not to be understood as compromise between different agricultural production systems. It rather must be understood as production system with a targeted, dynamic and continuous use and development of experiences which were made in the so-called conventional farming. In addition to natural scientific findings, impulses from organic farming are also taken up.

History

Integrated Pest Management can be seen as starting point for a holistic approach to agricultural production. Following the excessive use of crop protection chemicals, first steps in Integrated Pest Management (IPM) were taken in fruit production at the end of the 1950s. The concept was then further developed globally in all major crops. On the basis of results of the system-oriented IPM approach, models for Integrated Crop Management were developed. Initially, animal husbandry was not seen as part of such integrated approaches (Lütke Entrup et al., 1998 1).

In the years to follow, various national and regional initiatives and projects were formed. These include LEAF (Linking Environment And Farming) in the UK, FNL (Fördergemeinschaft Nachhaltige Landwirtschaft e.V.) in Germany, FARRE (Forum des Agriculteurs Responsables Respectueux de l'Environnement) in France, FILL (Fördergemeinschaft Integrierte Landbewirtschaftung Luxemburg) or OiB (Odling i Balans) in Sweden. However, there are few if any figures on the uptake of Integrated Farming in the major crops throughout Europe for example, leading to a recommendation by the European Economic and Social Committee in February 2014, that the EU should carry out an in-depth analysis of integrated production in Europe in order to obtain insights into the current situation and potential developments. There is evidence, however, that between 60 and 80% of pome, stone and soft fruits were grown, controlled and marketed according to "Integrated Production Guidelines" in 1999 already in Germany for example. In the UK, 22% of fresh fruit and vegetables are grown to Integrated Farm Management standards as recognised by the LEAF Marque.

Animal husbandry and Integrated Crop Management (ICM) often are just two branches of one agricultural enterprise. In modern agriculture, animal husbandry and crop production must be understood as interlinked sectors which cannot be looked at in isolation, as the context of agricultural systems leads to tight interdependencies. Uncoupling animal husbandry from arable production (too high stocking rates) is therefore not considered in accordance with the principles and objectives of Integrated Farming (Lütke Entrup et al., 1998 1). Accordingly, holistic concepts for Integrated Farming or Integrated Farm Management such as the EISA Integrated Farming Framework, and the concept of sustainable agriculture are increasingly developed, promoted and implemented at the global level.

Related to the 'sustainable intensification' of agriculture, an objective which in part is discussed controversially, efficiency of resource use becomes increasingly important today. Environmental impacts of agricultural production depend on the efficiency achieved when using natural resources and all other means of production. The input per kg of output, the output per kg of input, and the output achieved per hectare of land – a limited resource in the light of world population growth – are decisive figures for evaluating the efficiency and the environmental impact of agricultural systems. Efficiency parameters therefore offer important evidence how efficiency and environmental impacts of agriculture can be judged and where improvements can or must be made.

Against this background, documentation as well certification schemes and farm audits such as LEAF Marque in the UK and 33 other countries throughout the world become more and more important tools to evaluate – and further improve – agricultural practices. Even though being by far more product- or sector-oriented, SAI Platform principles and practices and GlobalGap for example pursue similar approaches.

Objectives

Continuous learning process in Integrated Farming

Integrated Farming is based on attention to detail, continuous improvement and managing all resources available.

Being bound to sustainable development, the underlying three dimensions "economic development", "social development" and "environmental protection" are thoroughly considered in the practical implementation of Integrated Farming. However, the need for profitability is a decisive prerequisite: To be sustainable, the system must be profitable, as profits generate the possibility to support all activities outlined in the (EISA Integrated Farming) IF Framework.

As a management and planning approach, Integrated Farming includes regular benchmarking of targets set against results achieved. The concept of the EISA Integrated Farming Framework for example has a clear focus on farmers' awareness of their own performance. By regularly benchmarking their performance, farmers become aware of achievements as well as deficiencies, and by paying attention to detail they can continuously work on improving the whole farming enterprise and their economic performance at the same time: According to findings in UK, reducing fertiliser and chemical inputs to amounts according to the demand of the crops allowed for cost savings in the range of £2,500 - £10,000 per year and per farm.

Prevalence

Following first developments in the 50s, various approaches to Integrated Pest Management, Integrated Crop Management, Integrated Production and Integrated Farming were developed worldwide (in Germany, Switzerland, US, Australia, and India, for example). As the implementation of the general concept of Integrated Farming and its individual components always should be handled according to the given site and situation instead of following strict rules and recipes, the concept is virtually applicable – and being used to various degrees – all over the world.

Criticism

It should be mentioned, however, that there are also critical voices from environmental organisations for example. That is in part due to the fact that there are European Organic Regulations such as (EC) No 834/2007 or the new draft from 2014 but no comparable regulations for Integrated Farming. Whereas organic farming and the "Bio-Siegel" in Germany for example are legally protected, EU Commission has not yet considered to start working on a comparable framework or blueprint for Integrated Farming. When products are marketed as "Controlled Integrated Produce", according control mechanisms and quality-labels are not based on national or European directives but are established and handled by private organisations and quality schemes such as LEAF Marque for example.

Agricultural Machinery

A German combine harvester by Claas

A British crop sprayer by Lite-Trac

Agricultural machinery is machinery used in farming or other agriculture. There are many types of such equipment, from hand tools and power tools to tractors and the countless kinds of farm implements that they tow or operate. Diverse arrays of equipment are used in both organic and nonorganic farming. Especially since the advent of mechanised agriculture, agricultural machinery is an indispensable part of how the world is fed.

History of the Machines

The Industrial Revolution

With the coming of the Industrial Revolution and the development of more complicated machines, farming methods took a great leap forward. Instead of harvesting grain by hand with a sharp blade, wheeled machines cut a continuous swath. Instead of threshing the grain by beating it with sticks, threshing machines separated the seeds from the heads and stalks. The first tractors appeared in the late 19th century.

Steam Power

Power for agricultural machinery was originally supplied by ox or other domesticated animals. With the invention of steam power came the portable engine, and later the traction engine, a multipurpose, mobile energy source that was the ground-crawling cousin to the steam locomotive. Agricultural steam engines took over the heavy pulling work of oxen, and were also equipped with a pulley that could power stationary machines via the use of a long belt. The steam-powered machines were low-powered by today's standards but, because of their size and their low gear ratios, they could provide a large drawbar pull. Their slow speed led farmers to comment that tractors had two speeds: "slow, and damn slow."

Internal Combustion Engines

The internal combustion engine; first the petrol engine, and later diesel engines; became the main source of power for the next generation of tractors. These engines also contributed to the development of the self-propelled, combined harvester and thresher, or combine harvester (also shortened to 'combine'). Instead of cutting the grain stalks and transporting them to a stationary threshing machine, these combines cut, threshed, and separated the grain while moving continuously through the field.

Types

A John Deere cotton harvester at work in a cotton field.

Combines might have taken the harvesting job away from tractors, but tractors still do the majority of work on a modern farm. They are used to push implements—machines that till the ground, plant seed, and perform other tasks.

From left to right: John Deere 7800 tractor with Houle slurry trailer, Case IH combine harvester, New Holland FX 25 forage harvester with corn head.

A New Holland TR85 combine harvester

Tillage implements prepare the soil for planting by loosening the soil and killing weeds or competing plants. The best-known is the plow, the ancient implement that was upgraded in 1838 by John Deere. Plows are now used less frequently in the U.S. than formerly, with offset disks used instead to turn over the soil, and chisels used to gain the depth needed to retain moisture.

The most common type of seeder is called a planter, and spaces seeds out equally in long rows, which are usually two to three feet apart. Some crops are planted by drills, which put out much more seed in rows less than a foot apart, blanketing the field with crops. Transplanters automate the task of transplanting seedlings to the field. With the widespread use of plastic mulch, plastic mulch layers, transplanters, and seeders lay down long rows of plastic, and plant through them automatically.

After planting, other implements can be used to cultivate weeds from between rows, or to spread fertilizer and pesticides. Hay balers can be used to tightly package grass or alfalfa into a storable form for the winter months.

Modern irrigation relies on machinery. Engines, pumps and other specialized gear provide water quickly and in high volumes to large areas of land. Similar types of equipment can be used to deliver fertilizers and pesticides.

Besides the tractor, other vehicles have been adapted for use in farming, including trucks, airplanes, and helicopters, such as for transporting crops and making equipment mobile, to aerial spraying and livestock herd management.

New Technology and the Future

The basic technology of agricultural machines has changed little in the last century. Though modern harvesters and planters may do a better job or be slightly tweaked from their predecessors, the US$250,000 combine of today still cuts, threshes, and separates grain in the same way it has always been done. However, technology is changing the way that humans operate the machines, as computer monitoring systems, GPS locators, and self-steer programs allow the most advanced tractors and implements to be more precise and less wasteful in the use of fuel, seed, or fertilizer. In the foreseeable future, there may be mass production of driverless tractors, which use GPS maps and electronic sensors.

Open Source Agricultural Equipment

Many farmers are upset by their inability to fix the new types of high-tech farm equipment. This is due mostly to companies using intellectual property law to prevent farmers from having the legal right to fix their equipment (or gain access to the information to allow them to do it). This has encouraged groups such as Open Source Ecology and Farm Hack to begin to make open source agricultural machinery. In addition on a smaller scale FarmBot and the RepRap open source 3D printer community has begun to make open-source farm tools available of increasing levels of sophistication. In October 2015 an exemption was added to the DMCA to allow inspection and modification of the software in cars and other vehicles including agricultural machinery.

Various Agricultural machinery

Disc Harrow

A modern Simba disc harrow

An Evers disc harrow

An offset (asymmetric) disc harrow

Disc harrow as part of a chisel plow by Case IH

A disc harrow is a farm implement that is used to till the soil where crops are to be planted. It is also used to chop up unwanted weeds or crop remainders. It consists of many carbon steel and sometimes the longer-lasting boron discs, which have many varying concavities and disc blade sizes and spacing (the choices of the later being determined by the final result required in a given soil type) and which are arranged into two sections ("offset disc harrow") or four sections ("tandem disc harrow"). When viewed from above, the four sections would appear to form an "X" which has been flattened to be wider than it is tall. The discs are also offset so that they are not parallel with the overall direction of the implement. This arrangement ensures that the discs will repeatedly slice any ground to which they are applied, in order to optimize the result. The concavity of the discs as well as their offset angle causes them to loosen and lift the soil that they cut.

A discer is an evolved form of a disc harrow, more suitable to Saskatchewan prairies, where it was developed in the 1940s. It does not leave ridging and it is lighter to pull, so it can be made bigger. After the 1980s their domination started to fade.

Name Variations

In various regions of the United States, farmers call these implements just discs (or disks), and they reserve the word *harrow* for the lighter types of harrow, such as chain and tooth harrows. Therefore, in these regions, the phrase "plowing, disking, and harrowing" refers to three separate tillage steps. This is not any official distinction but is how farmers tend to speak.

It is also common, at least in the United States, to consider disc plows to be a separate class of implement from discs (disc harrows). The first is a true plow, which does primary tillage and leaves behind a rough surface, whereas the second is a secondary tillage tool.

History

Before invention of the modern tractor, disc harrows typically consisted of two sections, which were horse-drawn and had no hydraulic power. These harrows were often adjustable so that the discs could be changed from their offset position. Straightening the discs allowed for transport without ripping up the ground; also, they were not as difficult to pull. Overuse of disc harrows in the High Plains of the United States in the early 20th century may have contributed to the "Dust Bowl".

Today

Modern disc harrows are tractor-driven and are raised either by a three-point lift or hydraulically by wheels. Some large ones even have side sections that can be raised vertically or that fold up to allow easier road transport or to provide better storage configurations.

Uses

Primary heavy duty disc harrows of 265 to 1000 lbs per disc are mainly used to break up virgin land, to chop material/residue, and to incorporate it into the top soil. Lighter secondary disc harrows help completely incorporate residue left by a primary disc harrow, eliminate clumps, and loosen the remaining packed soil. The notched disc blades chop up stover left from a previous crops, such as cornstalks. Disc harrows incorporate remaining residue into the top soil, promoting rapid the decay of the dead plant material. Applying fertilizer onto residue on the surface of the soil results in much of the applied nitrogen being tied up by residual plant material; therefore it is not available to germinating seeds. Disc harrows are also generally used prior to plowing in order to make the land easier to manage and work after plowing. Applying a disc harrow before plowing can also reduce clogging and allow more complete turning of the soil during plowing.

A disc harrow is the preferred method of incorporating both agricultural lime (either dolomitic or calcitic lime) and agricultural gypsum, and disc harrowing achieves a 50/50 mix with the soil when set correctly, thereby reducing acid saturation in the top soil and so promoting strong, healthy root development. Lime does not move in the soil, and this poses a critical challenge to sustainable zero-till farming, especially considering that chemical fertilizers are generally used by farmers around the world.

Offset Disc Harrow

The heavy duty disc with large diameter disc blades of 26", 28", 30", 32", 36", and 40", and with increased disc spacings of 10", 14", and 18" are the primary tillage tools that are used to break virgin ground, to incorporate residue into the soil in preparation for a ripper / subsoiler, and to break up a compacted soil in order to increase soil aeration and to promote soil permeability in lower levels of the soil profile. Prior to a planting operation, a secondary disc harrow with narrow disc spacing of 8", 9", and even 10" with disc sizes ranging from 20", 22", 24", to 26" can be used. Other similar secondary tillage tine implements or rotary harrows are also widely used. When choosing secondary tillage equipment, soil type as well as soil moisture content at the time must be considered. Lighter secondary disc harrows are primarily used to break down soil clods into smaller pieces. By so doing, water penetrates more easily into the soil, soil aeration is increased, and the activity of soil biota is enhanced; the final result is a seed bed that is suitable for planting.

Beaver Slide

Beaver Slide with Atlanta University in the background

Beaver Slide or Beavers' Slide was an African American slum area near Atlanta University documented as early as 1882. It was replaced by the University Homes public housing project in 1937, which was razed in 2008-9.

Charles Forrest Palmer, the man who organized the clearance of Beaver Slide and creation of University Homes, stated in his autobiographical book that Beaver Slide's name was due to James Beavers, Atlanta Chief of Police from 1911–1915, once observing the slum from a hillside, losing his footing and sliding down into the slum, thus: "Beavers' Slide".

In 1925 the area was targeted for a "cleanup" by city and university authorities.

The area was celebrated musically in the "Beaver Slide Rag" by Peg Leg Howell And His Gang, 1927.

It was finally razed to make for the University Homes public housing projects (William Augustus Edwards, architect), which opened in 1938.

University Homes was razed in 2008-9. As of January 2012 there has been no definitive announcement of what will be built on the land.

Roller (Agricultural Tool)

A 12 foot smooth roller

The roller is an agricultural tool used for flattening land or breaking up large clumps of soil, especially after ploughing. Typically, rollers are pulled by tractors or, prior to mechanisation, a team of animals such as horses or oxen. As well as for agricultural purposes, rollers are used on cricket pitches and residential lawn areas.

Flatter land makes subsequent weed control and harvesting easier, and rolling can help to reduce moisture loss from cultivated soil. On lawns, rolling levels the land for mowing and compacts the soil surface.

Rollers may be weighted in different ways. For many uses a heavy roller is used. These may consist of one or more cylinders made of thick steel, a thinner steel cylinder filled with concrete, or a cylinder filled with water. A water-filled roller has the advantage that the water may be drained out for lighter use or for transport. In frost-prone areas a water filled roller must be drained for winter storage to avoid breakage due to the expansion for water as it turns to ice.

Segmented Rollers

Cambridge roller-cultipacker

On tilled soil a one-piece roller has the disadvantage that when turning corners the outer end of the roller has to rotate much faster than the inner end, forcing one or both ends to skid. A one-piece roller turned on soft ground will skid up a heap of soil at the outer radius, leaving heaps, which is counter-productive. Rollers are often made in two or three sections to reduce this problem, and the

Cambridge roller overcomes it altogether by mounting many small segments onto one axle so that they can each rotate at local ground-speed.

A field after rolling with a Cambridge (or similar) roller

The surface of rollers may be smooth, or it may be textured to help break up soil or to groove the final surface to reduce scouring from rain. Each segment of a Cambridge roller has a rib around its edge for this purpose.

Rollers may be ganged, or combined with other equipment such as mowers.

Cricket Pitches

In cricket, rollers are used to make the pitch flat and less dangerous for batsmen.

Several size rollers have been used in the history of cricket, from light rollers that were used in the days of uncovered pitches and at some stages during the 1950s to make batting less easy, to the modern "heavy roller" universally used in top-class cricket today. Regulations permit a pitch only to be rolled at the commencement of each innings or day's play, but this has still had a massive influence on the game by eliminating the shooters that were ubiquitous on all but light soils before heavy rollers were used. Heavy rollers have sometimes been criticised for making batting too easy and for reducing the rate at which pitches dry out after rain in the cool English climate.

Lawn Rollers

112. Hand-roller.

Lawn roller

Lawn rollers are designed to even out or firm up the lawn surface, especially in climates where heaving causes the lawn to be lumpy. Heaving may result when the ground freezes and thaws many times over winter. Where this occurs, gardeners are advised to give the lawn a light rolling with a lawn roller in the spring. Clay or wet soils should not be rolled as they become compacted.

Threshing Board

Top view of a Spanish threshing board

Bottom view of a Spanish threshing board

A threshing board is an obsolete farm implement used to separate cereals from their straw; that is, to thresh. It is a thick board, made with a variety of slats, with a shape between rectangular and trapezoidal, with the frontal part somewhat narrower and curved upward (like a sled or sledge) and whose bottom is covered with lithic flakes or razor-like metal blades.

One form, once common in the Mediterranean Sea area, was "about three to four feet wide and six feet deep (these dimensions often vary, however), consisting of two or three wooden planks assembled to one another, of more than four inches wide, in which several hard and cutting flints crammed into the bottom part pull along over the grains. In the rear part there is a large ring nailed, that is used to tie the rope that pulls it and to which two horses are usually harnessed; and a person, sitting on the threshing board, drives it in circles over the cereal that is spread on the threshing floor. Should the person need more weight, he need only put some big stones over it."

The dimensions of threshing boards varied. In Spain, they could be up to approximately two metres in length and a metre and a half wide. There were also smaller threshing boards, as little about a metre-and-a-half long and a metre wide. The thickness of the slats of the threshing board is some five or six cm. Nonetheless, since threshing boards are nowadays custom made, made to order or made smaller as an adornment or souvenir, they may range from miniatures up to the sizes previously described.

The threshing board has been traditionally pulled by mules or by oxen over the grains spread on the threshing floor. As it was moved in circles over the harvest that was spread, the stone chips or blades cut the straw and the ear of wheat (which remained between the threshing board and the pebbles on the ground), thus separating the seed without damaging it. The threshed grain was then gathered and set to be cleaned by some means of winnowing.

Traditional Threshing Systems

Preparing sheafs to bring to the threshing floor

Traditional threshing with a threshing board in the Near East

Sweeping the threshing floor in order to pile up the seed

Cleaning the seed with a mechanical winnowing machine

Until the arrival of combine harvesters, which reap, thresh and clean grain in a single process, the traditional methods of threshing cereals and some legumes were those described by Pliny the Elder in his *Natural History*, with three variants: "The cereals are threshed in some places with the threshing board on the threshing floor; in others they are trampled by a train of horses, and in others they are beaten with flails".

In this manner, Pliny refers to the three traditional methods of threshing grain:

- Beating sheafs of grain against a crushing stone or a crushing lump of wood.

- Trampling grain spread on the threshing floor; the trampling would be done by a train of mules or oxen

- Threshing with flails, a type of traditional wooden tool with which one strikes the pile of grain until the seed is separated from the chaff.

Threshing with the Threshing Board

The threshing board is a historical form of threshing that can still be seen in some regions that practise a marginal agriculture. It is also somewhat preserved as an occasional folkloric and ceremonial practice, to commemorate traditional local customs.

For threshing with the threshing board, first one brings the baled stalks to the threshing floor. Some are stacked, waiting their turn, and others are untied and placed in a circle forming a pile of grain that is heated by the sun. Then, the farmers drag the threshing board over the stalks, first going several times around in circles, and then in figure-eights, and stirring the grain with a wooden pitchfork. Sometimes, this work was done with another kind of threshing implement: a *Plostellum punicum* (Latin; literally "Punic cart") or threshing cart, fitted with a group of rollers, each with metallic transverse razors. In this first stage, the straw is detached from the ear; much chaff and dirty dust remains, mixed with the edible grain. Every time that the work of dragging the threshing board is repeated, the grain is stirred again, moving more straw to the edge of the threshing floor. If too much grain is spread on the ground, it has to be raked and swept in order to make a round mound and, if possible, to remove as much straw as possible.

After turning the grain and straw upside down and leaving it to dry during a lunch break, the farmers carry out a second round of threshing in order to separate the last of the grain from the straw. Then, they use pitchforks, rakes and brooms to create a mound. A pair of oxen or mules pulls the threshing board by means of a chain or a strap fixed to a hook nailed in the front plank; donkeys are not used, because unlike mules and oxen they often defecate on the crops. The driver rides on the threshing board, both guiding the draft animals and increasing the weight of the threshing board. If the driver's weight is not enough, large stones are put on the board. In recent times, the animals are sometimes replaced with a tractor; because the driver no longer sits on the board, the weight of stones becomes more important. Children enjoy riding on the threshing board for fun, and the farmers allow it because their weight is useful, as long as the children are not too boisterous. During this process, if the stalks are excessively squashed, two large metal arcs are affixed to the back of the threshing board; these turn-up and give volume to the straws, behind the threshing.

After threshing is finished, to avoid mixing the dirty remnants with the clean, new stalks, the threshing floor must be cleaned first with a rake to move the heavier material, and then with several brooms (in the narrow sense: they are typically made from "broom shrub"—*Cytisus scoparius*—and are stronger than domestic brooms). Also the straw was accumulated carefully and stored, because it was a good fodder for livestock. The entire process of threshing generates a thin dust that soaks in through the respiratory system and sticks to the throat.

During the sweeping, the husks and the chaff are separated to one part of the threshing-floor, while the grain, still not entirely clean, was winnowed, either by traditional of winnowing with sieves, or by a mechanical winnowing machine.

Traditional threshing implements (including the threshing board) were gradually abandoned and replaced by modern combine harvesters. This change, of course, occurred in some areas before others. For instance, in Spain, it happened in the 1950s and 1960s. Until that time, threshing boards were made in certain particular towns and villages with specialised craftsmen. Whereas the woodworking involved is simple, even rough, the flintknapping and the inlaying of flakes into the bottom of board need specialised skills that were passed from father to son. The most famous Spanish town for this work was, doubtlessly, Cantalejo (Segovia), where the craftsmen who made threshing boards were known as *briqueros*.

History

Origins in the Neolithic and Copper Age

Obsidian trade in the 4th millennium BC

Engraved tablet from Kish, dating from 3350 BC, with representations of threshing boards on both sides

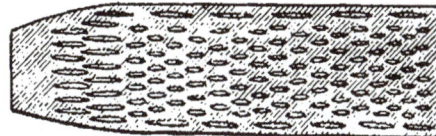

Small threshing board from Tunisia

Patricia C. Anderson (of *Centre d'Etudes Préhistoire, Antiquité et Moyen Age del CNRS*), discovered archaeological remains that demonstrate the existence of threshing boards at least 8,000 years old in the Near East and Balkans. The artefacts are lithic flakes and, above all obsidian or flint blades, recognizable through the type of microscopic wear that it has. Her work was completed by Jacques Chabot (of the *Centre interuniversitaire d'études sur les lettres, les arts et les traditions*, CELAT), who has studied Mitanni (northern Mesopotamia and Armenia). Both count among their specialties the study of microwear analysis, through which it is possible to take a particular piece of flint or obsidian (to take the most common examples) and determine the tasks for which it was used. Concretely, the cutting of cereals leaves a very characteristic *glossy* pattern of wear, owing to the presence of microscopic mineral particles (*phytoliths*) in the stalks of the plants. Therefore, scholars using controlled experimental replication studies and functional analysis with a scanning electron microscope are able to identify stone artefacts that were used as sickles or the teeth of threshing boards. The edge damage on the pieces used in threshing boards is distinct because, besides the *glossy* abrasion characteristic of cutting cereals, they have micro-scars from chipping, as a result of the blows of the threshing board against the rock surface of the threshing floor.

The most productive archaeological site is Aratashen, Armenia: a village occupied between 5000 and 3000 BC (Neolithic and Copper Age). The archaeological excavations have provided thousands of pieces from the knapping of obsidian (suggesting that Aratashen was a centre of production and trade of artefacts of that highly regarded stone); the rest of the archaeological record consists mainly of fragments of common pottery, ground stones, and other agricultural tools. Analysing a sample of 200 lithic flakes and blades, selected from the best-preserved pieces, it is possible to differentiate between those used in sickles and those used in threshing boards. The lithic blades of obsidian were knapped using highly developed and standardized methods, such as the use of a "pectoral crutch" with a copper point. Beginning at the headwaters of the river Euphrates, where this site is located, the craftsmen and peddlers sold their wares throughout the Middle East.

The threshing boards must have been important in the protohistory of Mesopotamia, since they already appear in some of the oldest written documents discovered: specifically, several sandstone tablets from the early town of Kish (Iraq), engraved with cuneiform pictograms, which could be the world's oldest surviving written record, dating to the middle of the 4th millennium BC (Early Uruk period). One of these tablets, preserved in the Ashmolean Museum of Oxford University, appears to have pictures of threshing boards on both faces, next to some numeric symbols and other pictograms. These presumed threshing boards (which might instead be sledges) have a shape similar to threshing carts that were used until recently in parts of the Middle East where non-industrial agriculture survived. Descriptions also appear in numerous cuneiform clay tablets as early as the third millennium BC.

Impression from a cylinder seal from Arslantepe-Malatya (Turkey), depicting a ritual thresh, dated to the third millennium BC

There are another representation, in this case without writing, in central Turkey. It is an impression of a cylinder seal from the archaeological site of Arslantepe-Malatya, which appeared near of the named «Temple B». The archaeological layers were dated to 3374 BC using dendrochronology. The stamp shows a figure seated on a threshing board, with a clear image of the lithic flakes inlaid in the bottom of the board. The main figure is sitting (possibly on a throne) under a dossal. In front is a driver or oxherd, and there are peasants with pitchforks nearby. According to M. A. Frangipane, that the seal may illustrate a religious scene:

It closely resembles another scene, painted on the walls of the same site (a ceremonial procession of a person of high rank, painted in an archaic lineal style in the colours red and black), although the current condition of the wall obscures the exact nature of the vehicle in which he is seated, it is indeed possible to see that it is pulled by a pair of oxen. Professor Sherratt interprets both scenes as presenting manifestations of civil or religious power. In that era, the threshing board was a sophisticated and expensive implement, built by specialized artisans, using pieces of flint or obsidian; in the case of Lower Mesopotamia, these were imported from far away: in the alluvial plateau of Sumer, as in all south of Mesopotamia, it was impossible to find stone, not even a pebble.

Furthermore, threshing required a threshing floor composed of natural rock, cobbles or, at least, very hard and compressed ground, located in a sunny place situated in a rise in the ground with a constant dry wind, and with a flat base that would not allow puddles to form when it rained. So, a threshing floor was not available to just anybody. It was expensive, as we can see from the biblical citations in the following section. Also, it required draft animals, expensive and difficult to drive (because this was not a matter of having them walk in a straight line). All this meant that a threshing implement required a large amount of harvested grain to pay off the expenditure. Thus, the rise of the threshing board turns on a distinctive, sophisticated system of powerful elites.

The discovery of a ceremonial sledge (perhaps a threshing board) with gold ornaments in the Tomb of Pu-Abi, one of the "royal tombs" of Ur, dated from the 3rd millennium BC, makes clear the underlying problem of distinguishing in the ancient representations between a true threshing board and a sledge (that is, an unwheeled vehicle for hauling freight). Although we know that the threshing board appears no later than the 4th millennium BC (as we can see in *Atarashen* and *Arslantepe-Malatya*), and we also know that the wheel was invented in Mesopotamia in the middle of that same millennium, still, the utilisation and spread of the wheel was not instantaneous. The sledges survived at least until the invention of the articulated axle, nearly 2000 BC. During this time, some vehicles were hybrids: sledges with wheels that could be dismantled to overcome obstacles by carrying it on shoulders or, simply, dragging it. Consequently—except in the case of Arslantepe, where the lithic flakes are clear visible—we cannot determine whether these representations are threshing boards or sledges for freight or for rites.

James Frazer compiled numerous ceremonies of harvest and thresh, that centered on a *Cereal Spirit*. From the Ancient Egyptian era to pre-industrial period, this spirit seems to have reside d in the first threshed sheaf or, sometimes, in the last one.

Biblical References

Any reference to the act of threshing or to the threshing board itself is incomplete without a reference to the *threshing floor*, because the work of threshing by necessity requires a threshing floor. The first biblical mention of the threshing floor is in Genesis 50:10. As such, it was not a shed, building, or any place covered with a roof and surrounded by walls, but a circular piece of ground from fifty to a hundred feet in diameter, in the open air, on elevated ground, and made smooth, hardy, and clean. Here the grain was threshed and winnowed.

In the Bible are found four modes of threshing grain (some of which modes are expounded in Isaiah 28:27-28):

- With a rod or flail. This was for small, delicate seeds such as flitches and cummin. It was also used when only small quantities of grain was to be threshed, or when it was necessary to conceal the operation from an enemy. Examples are found in Ruth 2:17, when Ruth "beat out" what she had gleaned during the day, and Judges 6:11, when Gideon "threshed (Hebrew, *chabat*, "beat out") wheat by the winepress to hide it from the Midianites."

- With the **charuts,** Hebrew for "threshing instrument." This is the threshing board or sledge under discussion in this article.

- With the **agalah,** Hebrew for "cart-wheel." This type of threshing instrument is probably referred to in Proverbs 22:26. The term *agalah* is supposed to have been the same as the **morag,** "threshing instrument," mentioned in 2Samuel 24:22, 1Chronicles 21:23, and Isaiah 41:15, though some make the *morag* and the *charuts* the same. This instrument is still known in Egypt by the name of **mowrej.** It consists of three or four heavy rollers of wood, iron, or stone, roughly made and joined together in a square frame, which is in the form of a sledge or drag. Rollers are said to be like barrels of an organ with their projections. Cylinders are parallel with each other and are stuck full of spikes having sharp square points. It is used the same way as the *charuts*. The driver sits on the machine, and with his weight helps to keep it down. (Authorities are not agreed as to the differences between the *charuts*, the *morag*, and the *agalah*.)

- By treading. The last mode is that of simply treading out the grain with a horse or an ox. Deuteronomy 25:4 is an example of this. The Egyptians used this rather inefficient mode of threshing with multiple oxen, and this mode is still in use in the East.

All four methods are discussed at length in the Talmud.

King David, by Pedro Berruguete

Two apparently coincident descriptive narratives are given in 2Samuel 24:10-25 and 1Chronicles 21:9-30 with regard to King David's purchase of the threshing floor on Mount Moriah (as well as Mount Moriah itself). In it, the Lord's directive to Gad, King David's prophet, was to instruct David to "rear an altar unto the Lord in the threshingfloor of Araunah the Jebusite" (2Samuel 24:18-19 and 1Chronicles 21:18-19) during the end of the three-day plague that was then ensuing upon David's people of Israel. In the 1Chronicles account, the purchase of the entire hilltop of Mount Moriah is the subject, which is why the purchase price (verse 25) is different from the 2 Samuel account.

This selection is highly significant for several reasons:

- It was selected by the Lord (not David) (2Samuel 24:18-19, 1Chronicles 21:18).

- Araunah's threshing floor was located on a hilltop called Mount Moriah. It is the same location of Abraham's attempted sacrifice of his son, Isaac (Genesis 22:2).

- The typology of the threshing floor is signified by Matthew 3:12 and Luke 3:17. It means that only the grain is gathered and admitted into the kingdom of God, while the chaff is cast into the unquenchable fire.

- On this hilltop was later built Solomon's Temple (2Chronicles 3:1), whose own typology heightens the significance of the threshing floor typology.

The selected place, Mount Moriah, was owned by Araunah (or, *Ornan*) the Jebusite, a rich villager who owned and operated the threshing floor on Mount Moriah. Araunah was awed by the visit of the king and offered his oxen to sacrifice and the threshing boards and other implements as wood for the sacrifice. Even so, David rejected any gift to come from a pagan and instead affirmed his intent to purchase first the oxen and threshing floor, then, later, the whole of Mount Moriah that contained it. Had David accepted Araunah's offer, it would have been Araunah's sacrifice instead of David's.

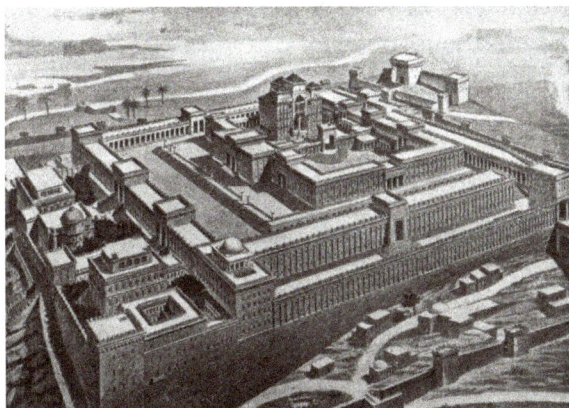

Solomon's Temple, built on the early theshing floor of Araunah the Jebusite

The references to the purchase are ostensively contradictory about the purchase price; it is 50 shekels of silver according to 2Samuel 24:24, and 600 shekels of gold according to 1Chronicles 21:25. However, this contradiction is only on the surface. The 50 shekels was the initial price to purchase the two oxen and the threshing floor, but the later price of 600 shekels was paid to purchase the property of Mount Moriah *in addition* to the oxen and threshing floor *contained within* and *earlier purchased*.

Generally, all threshing floors were located on hilltops, not to associate with the various gods, but to expose them to prevailing winds that assist in the process of winnowing. The wind adds to the typology of the threshing floor, as the wind is a type of the Holy Ghost.

The last biblical mention of threshing floors is in Matthew 3:12 and Luke 3:17. The term "floor" is the Greek *halon* which means threshing floor, and this fits the import of the two verses. The typology of the threshing floor on Mount Moriah, Solomon's Temple, and these two verses are significant.

As for the word "thresh," the last biblical mention is in 1Corinthians 9:10, again showing the typology of separating the chaff from the wheat.

Classical Greece and Rome

During the early history of Greece and Rome the threshing board was not used. Only after the development of commerce (with occurred in the 5th and 4th centuries in Greece and 2nd and 1st centuries in Rome) and the subsequent transmission of information from the near east that it became widely used. According to V.V. Struve, who cites, in part, verses of The Iliad, the Greeks of the 8th century BC threshed cereals by trampling them with oxen:

Carthage, which colonized the southeastern Iberian Peninsula in the 3rd century BC, had advanced agricultural technology, greatly superior to the Roman techniques of the time. Their methods astonished travellers such as Agathocles and Regulus, and were an inspiration for the writings of Varro and Pliny. One well-known Carthaginian agronomist, *Mago*, wrote a treatise that was translated into Latin by order of the Roman Senate. The ancient Romans describe Tunisia, today mainly desert, as a fertile landscape of olive groves and wheat fields. In Hispania, the Carthaginians are known to have introduced several new crops (mainly fruit trees) and some machines like the threshing board, either the version with stone-chips (*tribulum* in Latin) or the version with rollers (*threshing cart*, named in their honour *plostellum punicum* by the Romans).

In Rome, the threshing board had only economic significance, without the religious symbolism it took on in ancient Israel. The treatises of agriculture written by Roman experts as Cato, Varro, Columella and Pliny the Elder (quoted above), touch the topic of threshing. In chronological order:

- Cato: In the time of Cato the Elder—that is, the 2nd century BC—Rome was intensely connected with the conquered areas of Greece and Carthage, whose higher degree of agricultural development threatened Roman traditionalism. Cato's book *De Agricultura* was against exotic innovations such as the threshing board in its different variants, defending instead a traditional agricultural system based on manual labour. To some writers, Cato's ideas drove, indirectly, to the disintegration of republican society and even the imperial economy. Cato preferred threshing by trampling by mules or oxen. He doesn't expressly mention the threshing board, in spite of the fact it was already spreading through the empire. It is, then, "almost impossible to define, on the basis of Cato's report, when this or that implement or refinement came into use".

- Varro: Unlike Cato, Marcus Terentius Varro was not a man of action but a scholar, a πολιγραφοτάτω, in the 1st century BC. Varro, whose studies were wider than Cato's, tried to combine the cosmopolitan Greek outlook with Rome's provincial traditions. In his book of agricultural advice *Rerum Rusticarum de Agri Cultura* Varro only twice reflects the reality of his times by mentioning threshing boards. He advises, "None of the implements that can be produced in the plantation (farm) itself should be bought, as with almost all everything which is made from unfinished wood such as hampers, baskets, threshing boards, stakes, rakes…"; the inclination to self-sufficiency that he demonstrates here would later be harmful to Rome. Varro nonetheless shows himself more open to innovation that Cato: "To achieve an abundant and high-quality harvest, the stalks should be taken to the threshing floor without piling them up, so the grain is in the best condition, and the grain (should be) separated from the stalks on the threshing floor, a process which is done, among other ways, with a pair of mules and a threshing board. This is made with a wooden board (with its underside) equipped with stone-chips or saws of iron, which, with a plow in front or a large counterweight, is pulled by a pair of mules yoked together and thus separates the grain from the stalks…". That is, he explains in a very didactic way how the threshing boards works and the advantages of this innovative device. Next, he talks about the variant called *plostellum poenicum* (=*punicum*=Punic=Carthaginian), a threshing implement with rollers and metallic saws whose origin is, as we have already seen, Carthaginian, and which was used in Hispania (which had, in the past, been controlled by Carthage): "Another way to make it is by means of a cart with teethed rollers and bearings; this cart is named *plostellum punicum*, in which one can sit and move the device that is pulled by mules, as it is done in Hispania Citerior and other places."

- Columella (Lucius Junius Moderatus, beginning of Common Era - 60s): a native of Hispania Baetica; after finishing his military career, Columella worked managing large estates. This writer from Hispania brings a new note to this topic, writing, in this case, about threshing floors: "The threshing floor, if it is possible, must be placed in such way that it can be overseen by the master or by the foreman; the best is one that is cobbled, because not only allows that the cereal be quickly threshed, since the ground don't give way to the blows of hoofs and threshing boards, but also, these cereals, before being winnowed, are cleaner and lack the pebbles and little clods that always remaine in a threshing floor of pressed earth."

- Pliny: Pliny the Elder (23 - 79) only compiles what his predecessors had written, which we have already quoted.

Middle Ages

A Threshing board used as a door.

We will speak of the Middle Ages in a broad sense, without entering into great detail and focusing, essentially, on Western Europe because it is difficult to find any trustworthy documents about threshing boards in that era. The barbarian invasions of Europe had detrimental effects upon agriculture as well as upon the general economy, leading to the loss of many of the more advanced techniques, among them the threshing board, which was completely alien to Germanic tradition. The eastern Mediterranean areas, on the other hand, continued the use of the threshing board, passing into the Muslim culture, where it took deep root.

In the Iberian Peninsula, in the Visigothic kingdom and the Christian zone during the *Reconquista*, the threshing board was little known (although awareness of it never quite disappeared). The degradation extended not only to the economy, but also to the very sources that we have to study the period: scholars are confronted with a documentary void that is difficult to get around. It is certain that in Islamic Al-Andalus, the threshing board continued to be very popular, which led to the Christians recuperating the tradition as they advanced in the *Reconquista*. This act coincides with a generalized recovery in all of Europe. Economic prosperity began to return at the start of the 11th century; the experts speak of the increased area of tilled lands; the increased use of draft animals (first oxen, thanks to the frontal yoke, and later horses, thanks to harness collar); of the increase in metal tools and improved metalworking; of the appearance of the moldboard plough, often with wheels; and the increase in watermills. Livestock became a sign of progress: peasants, less dependent and more prosperous, became able to buy draft animals, and even plows. The peasants who had their own plow and one or two draft animals were a small elite, pampered by the feudal lord, who acquired a distinct status, that of *yeoman farmers*, quite distinct from *farm-hand labourer* whose only tool was their own arms. The existence of draft animals does not imply the diffusion of the threshing board in Western Europe, where the flail continued to be the preferred threshing implement. On the other hand, in Spain, the weight of Eastern tradition made the difference: Professor Julio Caro Baroja admits that in Spain the threshing board appears cited or represented in works of art. Concretely, he mentions some Romanesque reliefs in Beleña (Salamanca) and Campisábalos (Guadalajara), both from the 12th century. One may add a document written in 1265, in which a woman named Doña Mayor (widow of one Don Arnal, an ecclesiastical tax collector, and thus a person of good position), leaves to the diocese chapter of Salamanca her inheritance of *Valcuevo*, a farm in the municipality of Valverdón, Salamanca:

And I, Doña Mayor, leave on my death these two yokings of country state above referred to the Cathedral Chapter, well preserved with 76 bushels of wheat, 38 bushels of rye, and 38 bushels of barley to seed each yoking and with ploughshares, and with plows, and with ploughbeams, and with threshing boards and with all the accoutrements that a well-laid-out property should have.

These documents, at least testify for the presence of threshing boards, which, undoubtedly was continuous from then until recently in the Mediterranean basin. The rest is mere generalized speculation, given that traditional historiography centers on features more like of Western Europe. In any case, none of the consulted authors describe the threshing board as playing a relevant part in the progress of medieval agriculture. We must, then, join with the despair of French historian Georges Duby, in his complaint:

Through all that has been said, we can see how interesting it would be to measure the influence of technical progress on agricultural output. Nevertheless, we must renounce this hope. Before the end of the 12th century, the methods of seignorial administration were all quite primitive; they granted little importance to writing and even less to numbers. The documents are more deceptive than those of the Carolingian era.

Nowadays, numerous elements of traditional agriculture are being lost, and because of this various entities have been working to conserve or recover this cultural capital. Among these is an international interdisciplinary project called E.A.R.T.H., Early Agricultural Remnants and Technical Heritage. Participating countries include Bulgaria, Canada, France, Russia, Scotland, Spain, and the United States, in alphabetical order. Investigations center on broad archeological, documentary and ethnological aspects, related to diverse elements of traditional agriculture, threshing boards among them, in diverse countries, historical periods and societies.

Chronology of evidences referred in the article

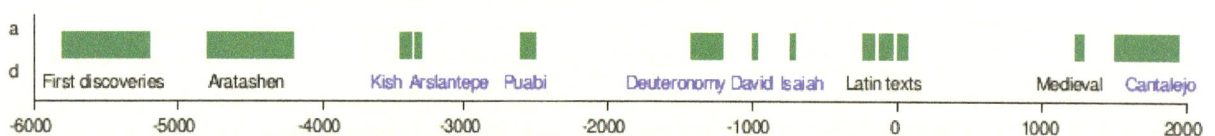

Craftsmen from Cantalejo

Cantalejo is located at the confluence of the Duraton River and the Cega River. Modern Cantalejo arose in the 11th century, although there are architectural remains that are much older, and forms part of Segovia province in Spain. It was apparently prosperous at the beginning but, lost its liberty in the 17th Century and became a jurisdictional lordship. There is no clear record of when the specialized craft of making threshing boards was introduced into Cantalejo, but all who have written on the topic indicate that it was probably during the 16th or 17th century.

The production of threshing boards appears to coincide with the arrival of foreign artisans, although this is speculation based on the fact that the artisans' form of speech includes aspects of many foreign languages, especially French. The makers of threshing boards and plows were called «Briqueros», a word of French origin that refers to the making of tinderboxes and matchlocks for guns, which in France were called *briquets*, literally in archaic French: *«Petite pièce d'acier don›t on se sert pour tirer du feu d'un caillou.»* (today, synonym of briquette or flammable matter, but also refers to a lighter) for the early firearms (Flintlocks, arquebuses, muskets...). It is therefore plausible that foreign gunsmiths introducing firearms in the Iberian Peninsula created an important colony in Cantalejo, although it is difficult to determine their origin

In any case, with time Cantalejo opted for peaceful and productive crafts such as the making of diverse agricultural implements, including threshing boards. By the 1950s, Cantalejo had reached 400 workshops producing more than 30,000 threshing boards each year; this suggests that more than half the population was dedicated to this occupation. The threshing boards were then distributed throughout the entire Meseta Central of Spain.

This article concentrates on threshing boards made only with flint chips, although there were threshing boards made with metal teeth or according to other, less common designs.

Making the Wooden Frame

At the end of summer, or in the autumn, work began with the selection of black pines, which they cut and carefully smoothed with a device called a *tronzador* until the trunks were formed into cylinders nearly two meters in length; these log sections were called *tozas*. They also prepared long, straight planks to serve as transverse headpieces. They carried the log sections to the sawmill, where they cut slats as wide as the log permitted (no less than 20 centimeters), of some five centimeters in thickness and with a curved shape (just like a ski) on the end that would eventually be in front. The planks were dried in the sun for several months, turned over every so often. The village during this period took on a peculiar appearance, as many buildings had their façades covered with drying planks. Afterwards, these were piled in *castillos*, crossing some slats with others in order to make the pile more stable.

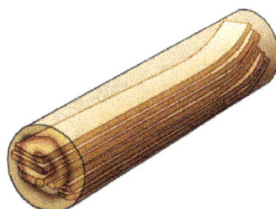

The pine *tozas*, from which the slats were cut

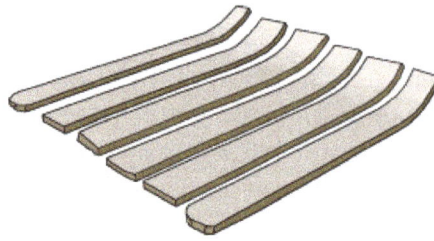

Slats aligned, before forming the threshing board

Chiseling the slat

Once the slat was in the right conditions, the next process was chiseling: using a hammer and chisel to prepare the slots (*ujeros*) for the chips of flint (*chinas* or lithic flakes). The chiseling was done with the slats front-on, guided by pencil-marks so the workman wouldn't err. Before beginning, the workman made sure that the slat had not warped since being cut, as that would make it unusable.

The next pass took place in the mechanical presses; the Spanish language term for these was *"cárceles"*, "jails". The three, four, or five planks had to be perfectly joined: spread with glue and pressed, using small reinforcement jigs, called *tasillos* (wooden cylinders glued and nailed with a mallet at the edges of a slat), and wedges. When the slats were well-aligned and fixed in place, the *cabezales* (headers), or crosspieces, were nailed in place with big nails known in Spain as *puntas de París* (although, at least in the 19th and 20th centuries, they came from Bilbao).

Once the basic structure of the threshing board is ready, it must be smoothed. This is first done by "working" it with an adze lengthwise, along the grain. Then the final finishing is done with various specialized carpenter's planes, on both top and bottom; first going across in a transverse direction, and later lengthwish.

The final phase of the work consisted of covering the junctions of the slats on the top side, which is done with thin strips at the front, tacked on with a board called the "front piece", and on the rest using long thin little boards (*tapajuntas* or stopgaps). On the front header they attached a strong hook for the *barzón*, or iron ring with a strap or a long rod for tying on the drafthorses or oxen.

Working the Stone Flakes

Various hammers traditionally found in the workshop of a *briquero*

To create the lithic flakes used to cover threshing boards, the *briqueros* in Cantalejo used a manufacturing technique similar to prehistoric methods of making tools, except that they used metal hammers rather than percussors made of stone, wood, or antler.

The raw materials preferred by these artisans was a whitish flint imported from the province of Guadalajara. When they had to repair threshing boards at home, if they did not have any other raw materials, they would use rounded river pebbles, made of homogeneous, high-grade quartzite, which they selected during their travels. The flint from Guadalajara was extracted from quarries in large blocks, which were split by hand with hammers of various sizes until the stone reached a size small enough to be comfortably held in the hand.

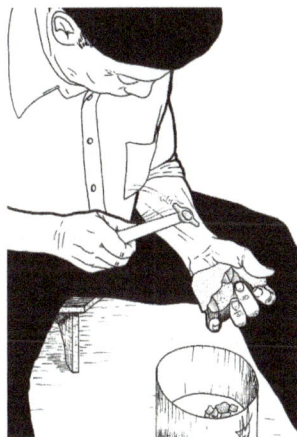

Briquero knapping flint

Knapping: Once manageable chunks of flint were obtained, knapping to obtain lithic flakes was performed using a very light hammer (called a *pickaxe*) with a narrow handle and a pointed head. Knapping was considered "men's work." To work quartzite pebbles, they used a hammer with a head that was rounded and slightly wider. During the process of removing stone flakes, they sometimes resorted to a mormal hammer to crack the stone and achieve perussion plains inaccessible with the pickaxe alone.

The *briquero* held the stone core in the left hand, protected with a piece of leather and with the palm upward, and struck rapid blows using a pick held in the right hand. The stone flakes would fall into the palm of the right hand, on top of the leather protector, which allowed the worker to evaluate them during fractions of a second: if they were acceptable, the *briquero* allowed them to fall into a tin; if not, he threw them into a reject pile. This pile was also where the *briquero* threw used-up stone cores —that is, stone blocks incapable of producing more chips; pieces of stone broken by accident; cortical flint flakes, useless fragments, and debris.

Woman pounding stone flakes into a threshing board

The working of pebbles was similar to the working of flint, except that with pebbles only the outside layer was chipped off. Thus, the pebbles were essentially "peeled" and discarded (unlike flint stones, whose interiors were worked until they were used up); using only the cortical flakes.

Covering the board with stone flakes was mainly the work of women called *enchifleras*. The task is monotonous and repetitive. Up to three thousand lithic flakes may be pounded into a large threshing board. In addition, it is necessary to sort the flakes: small ones in the front, medium-sized in the middle, and the largest on the sides and in the back. It is necessary to pound in each flake without damaging its sharp edge, although it was impossible to avoid leaving at least some small mark (a "spontaneous retouch" in technical terms). The tool used was a light hammer with a cylindrical head and flat or concave ends. The flakes are inserted into the cracks at their thickest part (technically, the percussion zone, that is, the proximal heel of the flake as it is struck off.)

Distribution

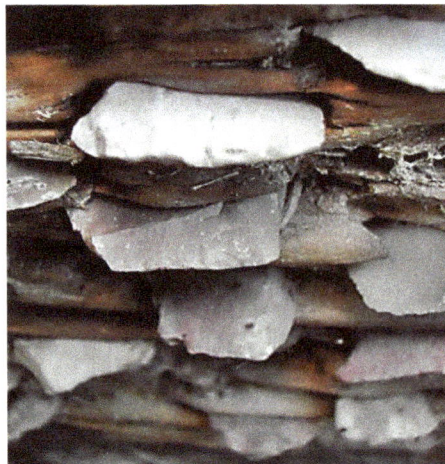

Detail of stone flakes: in foreground, a flake with a large chip; the others retain a sharp edge.

Threshing boards from Cantalejo captured nearly all the sales in Castile and León, Madrid, Castile-La Mancha, Aragon and Valencia. They sometimes also reached Andalusia y Cantabria. At first, artesans from Cantalejo travelled with large carts loaded with selected threshing boards, winnowing bellows, grain measures (of different traditional dry units: *celemín* is a wooden case with 4,6 L, *cuartilla* has 14 L, and *fanega* equivalent to 55.5 L...) and other implements for threshing or winnowing, which they peddled from town to town. They also carried flint chips, and tools and supplies for repairing damaged threshing boards and farming implements. In later times, they traveled by train to pre-arranged stations, and then in small trucks. They typically brought their entire families along; combined with their strange manner of speaking and unusual occupation, this gave the *briqueros* an air of mystery. They began selling threshing boards as soon as the threshing boards were complete, beginning in April and lasting until August. The *briqueros* would then return to their home town (the *Vilorio*) to celebrate the festivals of the Assumption of Mary (August 15) and Saint Roch (August 16) with their families.

Gacería

Gacería is a slang or argot used by makers and vendors of threshing boards in Cantalejo and some other parts of Spain. It is not a technical vocabulary, but rather a code made up of a small group of words that allows the speakers to communicate freely in the presence of strangers without others understanding the content of the conversation.

Gacería was purely verbal, colloquial, and associated with the selling of threshing boards; as a result, it largely disappeared with the mechanization of agriculture. Nevertheless, a number of studies have attempted to record its varied aspects. There are many doubts regarding the origin of the words that make up the vocabulary of *Gacería*, including the word *Gacería* itself, which may derive from the Basque word *gazo,* which means "ugly" or "good-for-nothing". The most commonly accepted opinion is that most of the words derive from French, with additions from other languages including Latin, Basque, Arabic, German, and even Caló. What is certain is that the makers and vendors of threshing boards took words from any area they visited regularly, creating a linguistic mishmash.

Related Threshing Implements

In general, the term "threshing board" is used to refer to all the different variants of this primitive implement. Technically, we should distinguish at least the two main types of threshing boards: the "threshing sledge," which is the subject of this article, and the "threshing cart."

The "threshing sledge" is the most common type. As its name indicates, it is dragged over the ripe grain, and it threshes using cutting pieces made of stone or metal. This is what is referred to in Hebrew as *morag* (מורג) and in Arabic as *mowrej*. Strictly speaking, the threshing boards of the Middle East have characteristics that render them easily distinguishable from those found in Europe. On the Iberian peninsula, cutting blades found on the bottom part of the threshing board are arranged on end and in rows roughly parallel to the direction of threshing. In contrast, the *mogag* and *mowrej* found to the Middle East have circular holes (made with a special short, wide drill) into which are pressed small round, semicircular stones with sharp ridges.

As mentioned previously, not all threshing boards are equipped with stones: some have metal knife blades embedded along the full length of the threshing board, and others have smaller blades encrusted here and there. Threshing sledges with metal knives usually have a few small wheels (four to six, depending on the size) with eccentric axes. These wheels protect the blades. They also make the threshing board wobble, causing parts of the board to rise and fall at random in an oscillating motion that improves the effectiveness of the threshing.

| Threshing sledge Palestine, 1937 | Threshing cart (*plostellum punicum*) in Heliopolis, Egypt in 1884 |

A second model, which the classic sources refer to as *plostellum punicum* (literally, the "Punic cart"), ought to be called "threshing cart". Although the Carthaginians, heirs of the Phoenicians, brought this model to the Western Mediterranean, this implement was known at least since the second millennium BC, appearing in the Babylonian texts with a name that we can transliterate as *gīš-bad*. Both variants continued to be used well into the 20th century in Europe, and continue to be used in the regions where agriculture has not been mechanized and industrialized. Museums and collectors in Spain retain some threshing carts, which were once highly prized in areas such as the province of Zamora, where they were used to thresh garbanzos.

References

- Pearce, J.M.(2015). Applications of Open Source 3-D Printing on Small Farms. *Organic Farming* 1(1), 19-35. DOI: 10.12924/of2014.01010019

- This article draws heavily on the corresponding article Trillo (agricultura) in the Spanish-language Wikipedia, accessed in the version of 20 November 2006.

- The big threshing boards were intended for a "yoke" or pair of draft animals (oxen or mules), whereas the smaller ones were indended to be drawn by a single animal

- Lucas Varela, Antonio (2002). Cerramícalo (in Spanish). Fundación Hernández Puertas de Alaeojs (Valladolid) - *Madrid. ISBN 84-607-4578-3.*)

- Sherratt, Andrew (2005). "ArchAtlas, an electronic atlas of archaeology Animal traction and the transformation of Europe". P. Pétrequin. Proceedings of the Frasnois Conference. *Sheffield, England* (p. 19–21).

- Hamblin, Dora Jane (1973). The First Cities. Time-Life Inc. *ISBN 0-8094-1301-9.* p.90. Cite error: Invalid <ref> tag; name "hamblin" defined multiple times with different content (see the help page).

- *Frazer, James George* (1922, reprint of 1995). *The Golden Bough.* Touchstone edition. *ISBN 0-684-82630-5.*

Check date values in: |date= (help) (p. 488 and followings)

- Struve, V. V. (1979). Historia de la Antigua Grecia (in Spanish) (Third ed.). *Madrid*: Akal Editor. *ISBN 84-7339-190-X.* p. 115

- *Duby, Georges* (1974). The early growth of the European economy. Warriors and Peasants from the Seventh to the Twelfth Century. Ithaca, New York: Cornell University Press. *ISBN 0-8014-9169-X.* .

- *Caro Baroja Julio* (1983). Tecnología popular española. Editorial Nacional, Colección Artes del tiempo y del espacio, *Madrid. ISBN 84-276-0588-9.* p. 98

- The activity is rhythmic and very rapid, too fast for an inexperinced observer to determine whether a chip is adequate or not to use in a threshing board -- whether it is too small, or too large, or has the proper shape.

- Cuesta Polo, Marciano, coord. (1993). Glosario de Gacería (in Spanish). *Ayuntamiento de Cantalejo, Segovia.* CS1 maint: Multiple names: authors list (link) p. 5

Introduction to Farming

This chapter is an introduction to farming incorporating all the major aspects related to the area of study. The chapter unfolds its crucial aspects in a critical yet systematic manner. It will not only give a detailed account of farming but it will also, discuss at length the related dimensions of it like the various types of farms, equipments used for farming across the globe, etc.

Farm

Farmland in the USA. The round fields are due to the use of center pivot irrigation

A farm is an area of land that is devoted primarily to agricultural processes with the primary objective of producing food and other crops; it is the basic facility in food production. The name is used for specialised units such as arable farms, vegetable farms, fruit farms, dairy, pig and poultry farms, and land used for the production of natural fibres, biofuel and other commodities. It includes ranches, feedlots, orchards, plantations and estates, smallholdings and hobby farms, and includes the farmhouse and agricultural buildings as well as the land. In modern times the term has been extended so as to include such industrial operations as wind farms and fish farms, both of which can operate on land or sea.

Farming originated independently in different parts of the world as hunter gatherer societies transitioned to food production rather than food capture. It may have started about 12,000 years ago with the domestication of livestock in the Fertile Crescent in western Asia, soon to be followed by the cultivation of crops. Modern units tend to specialise in the crops or livestock best suited to the region, with their finished products being sold for the retail market or for further processing, with farm products being traded around the world.

Typical plan of a mediaeval English manor, showing the use of field strips

Modern farms in developed countries are highly mechanized. In the United States, livestock may be raised on rangeland and finished in feedlots and the mechanisation of crop production has brought about a great decrease in the number of agricultural workers needed. In Europe, traditional family farms are giving way to larger production units. In Australia, some farms are very large because the land is unable to support a high stocking density of livestock because of climatic conditions. In less developed countries, small farms are the norm, and the majority of rural residents are subsistence farmers, feeding their families and selling any surplus products in the local market.

Etymology

A farmer harvesting crops with mule-drawn wagon, 1920s, Iowa, USA

The word in the sense of an agricultural land-holding derives from the verb "to farm" a revenue source, whether taxes, customs, rents of a group of manors or simply to hold an individual manor by the feudal land tenure of "fee farm". The word is from the medieval Latin noun *firma*, also the source of the French word *ferme*, meaning a fixed agreement, contract, from the classical Latin adjective *firmus* meaning strong, stout, firm. As in the medieval age virtually all manors were engaged in the business of agriculture, which was their principal revenue source, so to hold a manor by the tenure of "fee farm" became synonymous with the practice of agriculture itself.

History

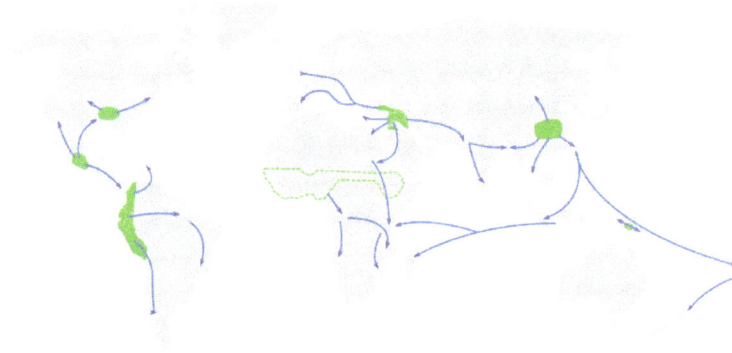

Map of the world showing approximate centers of origin of agriculture and its spread in prehistory: the Fertile Crescent (11,000 BP), the Yangtze and Yellow River basins (9,000 BP), and the New Guinea Highlands (9,000–6,000 BP), Central Mexico (5,000–4,000 BP), Northern South America (5,000–4,000 BP), sub-Saharan Africa (5,000–4,000 BP, exact location unknown), eastern North America (4,000–3,000 BP).

Farming has been innovated at multiple different points and places in human history. The transition from hunter-gatherer to settled, agricultural societies is called the Neolithic Revolution and first began around 12,000 years ago, near the beginning of the geological epoch of the Holocene around 12,000 years ago. It was the world's first historically verifiable revolution in agriculture. Subsequent step-changes in human farming practices were provoked by the British Agricultural Revolution in the 18th century, and the Green Revolution of the second half of the 20th century. Farming spread from the Middle East to Europe and by 4,000 BC people that lived in the central part of Europe were using oxen to pull plows and wagons.

Types of Farm

A farm may be owned and operated by a single individual, family, community, corporation or a company, may produce one or many types of produce, and can be a holding of any size from a fraction of a hectare to several thousand hectares.

A farm may operate under a monoculture system or with a variety of cereal or arable crops, which may be separate from or combined with raising livestock. Specialist farms are often denoted as such, thus a dairy farm, fish farm, poultry farm or mink farm.

Some farms may not use the word at all, hence vineyard (grapes), orchard (nuts and other fruit), market garden or "truck farm" (vegetables and flowers). Some farms may be denoted by their topographical location, such as a hill farm, while large estates growing cash crops such as cotton or coffee may be called plantations.

Many other terms are used to describe farms to denote their methods of production, as in collective, corporate, intensive, organic or vertical.

Other farms may primarily exist for research or education, such as an ant farm, and since farming is synonymous with mass production, the word "farm" may be used to describe wind power generation or puppy farm.

Specialized Farms

Dairy Farm

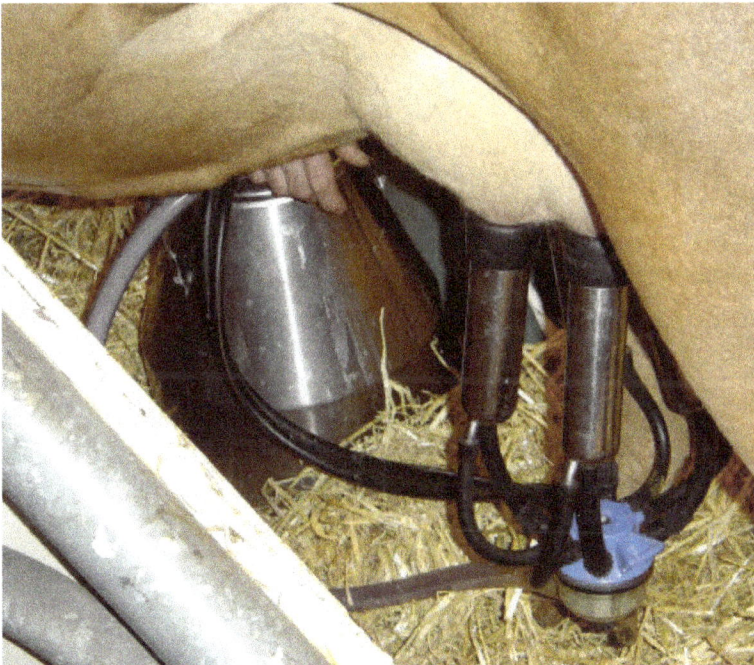

A milking machine in action

Dairy farming is a class of agriculture, where female cattle, goats, or other mammals are raised for their milk, which may be either processed on-site or transported to a dairy for processing and eventual retail sale. There are many breeds of cattle that can be milked some of the best producing ones include Holstein, Norwegian Red, Kostroma, Brown Swiss, and more.

In most Western countries, a centralized dairy facility processes milk and dairy products, such as cream, butter, and cheese. In the United States, these dairies are usually local companies, while in the southern hemisphere facilities may be run by very large nationwide or trans-national corporations (such as Fonterra).

Dairy farms generally sell male calves for veal meat, as dairy breeds are not normally satisfactory for commercial beef production. Many dairy farms also grow their own feed, typically including corn, alfalfa, and hay. This is fed directly to the cows, or stored as silage for use during the winter season. Additional dietary supplements are added to the feed to improve milk production.

Poultry farm

Poultry farming

Poultry farms are devoted to raising chickens (egg layers or broilers), turkeys, ducks, and other fowl, generally for meat or eggs.

Pig Farm

A pig farm is one that specializes in raising pigs or hogs for bacon, ham and other pork products and may be free range, intensive, or both.

Prison Farm

Prison farms are farms which serve as prisons for people sentenced to hard labor by a court. On prison farms inmates run the important tasks of a farm and producing crops.

Ownership

Farm control and ownership has traditionally been a key indicator of status and power, especially in Medieval European agrarian societies. The distribution of farm ownership has historically been closely linked to form of government. Medieval feudalism was essentially a system that centralized control of farmland, control of farm labor and political power, while the early American democracy, in which land ownership was a prerequisite for voting rights, was built on relatively easy paths to individual farm ownership. However, the gradual modernization and mechanization of farming, which greatly increases both the efficiency and capital requirements of farming, has led to increasingly large farms. This has usually been accompanied by the decoupling of political power from farm ownership.

Forms of Ownership

In some societies (especially socialist and communist), collective farming is the norm, with either government ownership of the land or common ownership by a local group. Especially in societies without widespread industrialized farming, tenant farming and sharecropping are common; farmers either pay landowners for the right to use farmland or give up a portion of the crops.

Farms Around the World

Americas

Farming near Klingerstown, Pennsylvania

The land and buildings of a farm are called the "farmstead". Enterprises where livestock are raised on rangeland are called *ranches*. Where livestock are raised in confinement on feed produced elsewhere, the term *feedlot* is usually used.

In 1910 there were 6,406,000 farms and 10,174,000 family workers; In 2000 there were only 2,172,000 farms and 2,062,300 family workers. The share of U.S. farms operated by women has risen steadily over recent decades, from 5 percent in 1978 to 14 percent by 2007.

A typical North American grain farm with farmstead in Ontario, Canada

In the United States, there are over three million migrant and seasonal farmworkers; 72% are foreign-born, 78% are male, they have an average age of 36 and average education of 8 years. Farmworkers make an average hourly rate of $9–10 per hour, compared to an average of over $18 per hour for nonfarm labor. Their average family income is under $20,000 and 23% live in families with incomes below the federal poverty level. One-half of all farmworker families earn less than $10,000 per year, which is significantly below the 2005 U.S. poverty level of $19,874 for a family of four.

In 2007, corn acres are expected to increase by 15% because of the high demand for ethanol, both in and outside of the U.S. Producers are expecting to plant 90.5 million acres (366,000 km²) of corn, making it the largest corn crop since 1944.

Asia

Farmlands in Hebei province, China

Pakistan

According to the World Bank, "most empirical evidence indicates that land productivity on large farms in Pakistan is lower than that of small farms, holding other factors constant." Small farmers have "higher net returns per hectare" than large farms, according to farm household income data.

Nepal

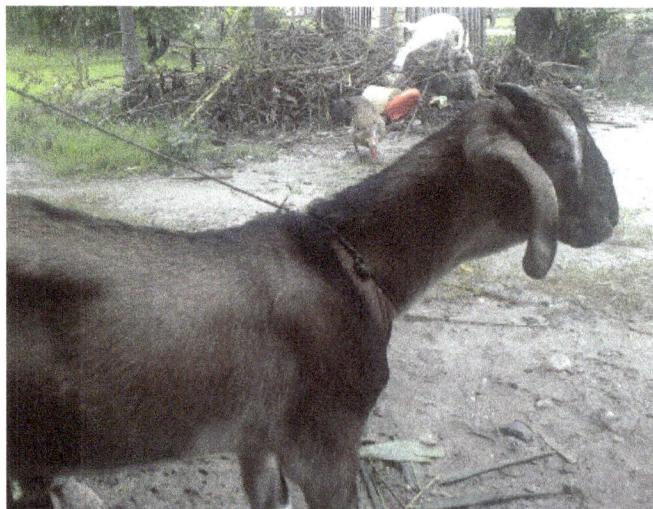
Goat found in Nepal

Nepal is an agricultural country and about 80% of the total population are engaged in farming. Rice is mainly produced in Nepal along with fruits like apples. Dairy farming and poultry farming are also growing in Nepal.

Australia

Cows grazing on a farm in Victoria, Australia

Farming is a significant economic sector in Australia. A farm is an area of land used for primary production which will include buildings.

According to the UN, "green agriculture directs a greater share of total farming input expenditures towards the purchase of locally sourced inputs (e.g. labour and organic fertilisers) and a local multiplier effect is expected to kick in. Overall, green farming practices tend to require more labour inputs than conventional farming (e.g. from comparable levels to as much as 30 per cent more) (FAO 2007 and European Commission 2010), creating jobs in rural areas and a higher return on labour inputs."

Where most of the income is from some other employment, and the farm is really an expanded residence, the term *hobby farm* is common. This will allow sufficient size for recreational use but be very unlikely to produce sufficient income to be self-sustaining. Hobby farms are commonly around 2 hectares (4.9 acres) but may be much larger depending upon land prices (which vary regionally).

Often very small farms used for intensive primary production are referred to by the specialization they are being used for, such as a dairy rather than a dairy farm, a piggery, a market garden, etc. This also applies to feedlots, which are specifically developed to a single purpose and are often not able to be used for more general purpose (mixed) farming practices.

In remote areas farms can become quite large. As with *estates* in England, there is no defined size or method of operation at which a large farm becomes a station.

Europe

Traditional Dutch farmhouse

In the UK, *farm* as an agricultural unit, always denotes the area of pasture and other fields together with its farmhouse, farmyard and outbuildings. Large farms, or groups of farms under the same ownership, may be called an estate. Conversely, a small farm surrounding the owner's dwelling is called a smallholding and is generally focused on self-sufficiency with only the surplus being sold.

Farm Equipment

Farm equipment has evolved over the centuries from simple hand tools such as the hoe, through ox- or horse-drawn equipment such as the plough and harrow, to the modern highly-technical machinery such as the tractor, baler and combine harvester replacing what was a highly labour-intensive occupation before the Industrial revolution. Today much of the farm equipment used on both small and large farms is automated (e.g. using satellite guided farming).

As new types of high-tech farm equipment have become inaccessible to farmers that historically fixed their own equipment, Wired reports there is a growing backlash, due mostly to companies using intellectual property law to prevent farmers from having the legal right to fix their equipment (or gain access to the information to allow them to do it). This has encouraged groups such as Open Source Ecology and Farm Hack to begin to make open source hardware for agricultural machinery. In addition on a smaller scale Farmbot and the RepRap open source 3D printer community has begun to make open-source farm tools available of increasing levels of sophistication.

Various Specialized Farms

Dairy Farming

Dairy farming is a class of agriculture for long-term production of milk, which is processed (either on the farm or at a dairy plant, either of which may be called a dairy) for eventual sale of a dairy product.

A rotary milking parlor at a modern dairy facility, located in Germany

Common species

Although any mammal can produce milk, commercial dairy farms are typically one-species enterprises. In developed countries, dairy farms typically consist of high producing dairy cows. Other species used in commercial dairy farming include goats, sheep, and camels. In Italy, donkey dairies are growing in popularity to produce an alternative milk source for human infants.

A dairy farm on the banks of the Columbia River in Clark County, Washington (May 1973).

History

Dairy farming has been part of agriculture for thousands of years. Historically it has been one part of small, diverse farms. In the last century or so larger farms doing only dairy production have emerged. Large scale dairy farming is only viable where either a large amount of milk is required for production of more durable dairy products such as cheese, butter, etc. or there is a substantial market of people with cash to buy milk, but no cows of their own.

Hand Milking

Woman hand milking a cow.

Centralized dairy farming as we understand it primarily developed around villages and cities, where residents were unable to have cows of their own due to a lack of grazing land. Near the town, farmers could make some extra money on the side by having additional animals and selling the milk in town. The dairy farmers would fill barrels with milk in the morning and bring it to market on a wagon. Until the late 19th century, the milking of the cow was done by hand. In the United States, several large dairy operations existed in some northeastern states and in the west, that involved as many as several hundred cows, but an individual milker could not be expected to milk more than a dozen cows a day. Smaller operations predominated.

For most herds, milking took place indoors twice a day, in a barn with the cattle tied by the neck with ropes or held in place by stanchions. Feeding could occur simultaneously with milking in the barn, although most dairy cattle were pastured during the day between milkings. Such examples of this method of dairy farming are difficult to locate, but some are preserved as a historic site for a glimpse into the days gone by. One such instance that is open for this is at Point Reyes National Seashore.

Dairy farming has been part of agriculture for thousands of years. Historically it has been one part of small, diverse farms. In the last century or so larger farms doing only dairy production have emerged. Large scale dairy farming is only viable where either a large amount of milk is required for production of more durable dairy products such as cheese, butter, etc. or there is a substantial market of people with cash to buy milk, but no cows of their own.

Vacuum Bucket Milking

The first milking machines were an extension of the traditional milking pail. The early milker device fit on top of a regular milk pail and sat on the floor under the cow. Following each cow being milked, the bucket would be dumped into a holding tank. These were introduced in the early 20th century.

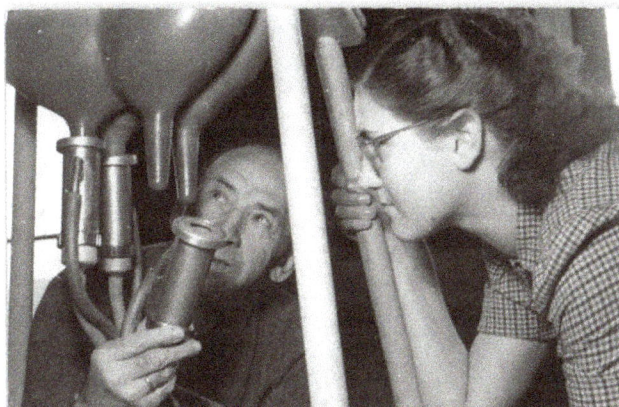

Demonstration of a new Soviet milker device. East Germany, 1952

This developed into the Surge hanging milker. Prior to milking a cow, a large wide leather strap called a surcingle was put around the cow, across the cow's lower back. The milker device and collection tank hung underneath the cow from the strap. This innovation allowed the cow to move around naturally during the milking process rather than having to stand perfectly still over a bucket on the floor.

Milking Pipeline

The next innovation in automatic milking was the milk pipeline, introduced in the late 20th century. This uses a permanent milk-return pipe and a second vacuum pipe that encircles the barn or milking parlor above the rows of cows, with quick-seal entry ports above each cow. By eliminating the need for the milk container, the milking device shrank in size and weight to the point where it could hang under the cow, held up only by the sucking force of the milker nipples on the cow's udder. The milk is pulled up into the milk-return pipe by the vacuum system, and then flows by gravity to the milkhouse vacuum-breaker that puts the milk in the storage tank. The pipeline system greatly reduced the physical labor of milking since the farmer no longer needed to carry around huge heavy buckets of milk from each cow.

The pipeline allowed barn length to keep increasing and expanding, but after a point farmers started to milk the cows in large groups, filling the barn with one-half to one-third of the herd, milking the animals, and then emptying and refilling the barn. As herd sizes continued to increase, this evolved into the more efficient milking parlor.

Milking Parlors

Innovation in milking focused on mechanizing the milking parlor (known in Australia and New Zealand as a *milking shed*) to maximize the number of cows per operator which streamlined the milking process to permit cows to be milked as if on an assembly line, and to reduce physical stresses on the farmer by putting the cows on a platform slightly above the person milking the cows to eliminate having to constantly bend over. Many older and smaller farms still have tie-stall or stanchion barns, but worldwide a majority of commercial farms have parlors.

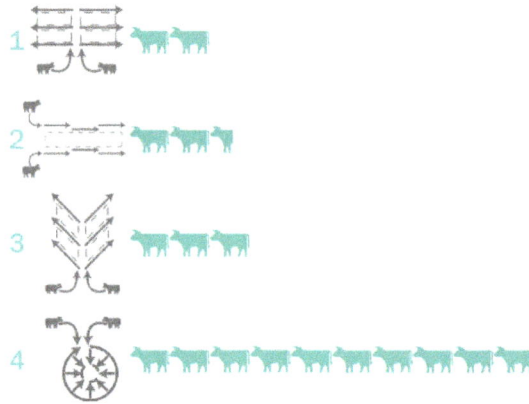

Efficiency of four different milking parlors. 1=Bali-Style 50 cows/h. 2=Swingover 60 cows/h.
3=Herringbone 75 cows/h. 4=Rotary 250 cows/h.

Herringbone and Parallel Parlors

In herringbone and parallel parlors, the milker generally milks one row at a time. The milker will move a row of cows from the holding yard into the milking parlor, and milk each cow in that row. Once all of the milking machines have been removed from the milked row, the milker releases the cows to their feed. A new group of cows is then loaded into the now vacant side and the process repeats until all cows are milked. Depending on the size of the milking parlor, which normally is the bottleneck, these rows of cows can range from four to sixty at a time.

Rotary Parlors

Rotary milking parlor

In rotary parlors, the cows are loaded one at a time onto the platform as it rotates. The milker stands near the entry to the parlor and puts the cups on the cows as they move past. By the time the platform has completed almost a full rotation, another milker or a machine removes the cups and the cow steps backwards off the platform and then walks to its feed. Rotary cowsheds, as they are called in New Zealand, started in the 1980s but are expensive compared to Herringbone cowshed - the older New Zealand norm.

Automatic Milker Take-Off

It can be harmful to an animal for it to be over-milked past the point where the udder has stopped releasing milk. Consequently, the milking process involves not just applying the milker, but also monitoring the process to determine when the animal has been *milked out* and the milker should be removed. While parlor operations allowed a farmer to milk many more animals much more quickly, it also increased the number of animals to be monitored simultaneously by the farmer. The automatic take-off system was developed to remove the milker from the cow when the milk flow reaches a preset level, relieving the farmer of the duties of carefully watching over 20 or more animals being milked at the same time.

Fully Automated Robotic Milking

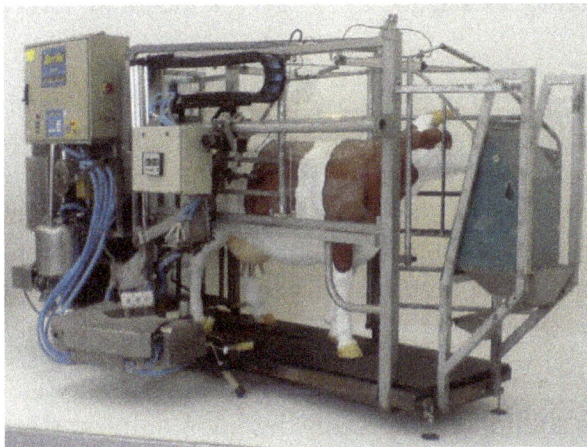

An automatic milking system unit as an exhibit at a museum

In the 1980s and 1990s, robotic milking systems were developed and introduced (principally in the EU). Thousands of these systems are now in routine operation. In these systems the cow has a high degree of autonomy to choose her time of milking freely during the day (some alternatives may apply, depending on cow-traffic solution used at a farm level). These systems are generally limited to intensively managed systems although research continues to match them to the requirements of grazing cattle and to develop sensors to detect animal health and fertility automatically. Every time the cow enters the milking unit she is fed concentrates and her collar is scanned to record production data.

History of Milk Preservation Methods

Cool temperature has been the main method by which milk freshness has been extended. When windmills and well pumps were invented, one of their first uses on the farm, besides providing water for animals themselves, was for cooling milk, to extend its storage life, until it would be transported to the town market.

The naturally cold underground water would be continuously pumped into a cooling tub or vat. Tall, ten-gallon metal containers filled with freshly obtained milk, which is naturally warm, were placed in this cooling bath. This method of milk cooling was popular before the arrival of electricity and refrigeration.

Refrigeration

When refrigeration first arrived (the 19th century) the equipment was initially used to cool cans of milk, which were filled by hand milking. These cans were placed into a cooled water bath to remove heat and keep them cool until they were able to be transported to a collection facility. As more automated methods were developed for harvesting milk, hand milking was replaced and, as a result, the milk can was replaced by a bulk milk cooler. 'Ice banks' were the first type of bulk milk cooler. This was a double wall vessel with evaporator coils and water located between the walls at the bottom and sides of the tank. A small refrigeration compressor was used to remove heat from the evaporator coils. Ice eventually builds up around the coils, until it reaches a thickness of about three inches surrounding each pipe, and the cooling system shuts off. When the milking operation starts, only the milk agitator and the water circulation pump, which flows water across the ice and the steel walls of the tank, are needed to reduce the incoming milk to a temperature below 5 degrees.

This cooling method worked well for smaller dairies, however was fairly inefficient and was unable to meet the increasingly higher cooling demand of larger milking parlors. In the mid-1950s direct expansion refrigeration was first applied directly to the bulk milk cooler. This type of cooling utilizes an evaporator built directly into the inner wall of the storage tank to remove heat from the milk. Direct expansion is able to cool milk at a much faster rate than early ice bank type coolers and is still the primary method for bulk tank cooling today on small to medium-sized operations.

Another device which has contributed significantly to milk quality is the plate heat exchanger (PHE). This device utilizes a number of specially designed stainless steel plates with small spaces between them. Milk is passed between every other set of plates with water being passed between the balance of the plates to remove heat from the milk. This method of cooling can remove large amounts of heat from the milk in a very short time, thus drastically slowing bacteria growth and thereby improving milk quality. Ground water is the most common source of cooling medium for this device. Dairy cows consume approximately 3 gallons of water for every gallon of milk production and prefer to drink slightly warm water as opposed to cold ground water. For this reason, PHE's can result in drastically improved milk quality, reduced operating costs for the dairymen by reducing the refrigeration load on his bulk milk cooler, and increased milk production by supplying the cows with a source of fresh warm water.

Plate heat exchangers have also evolved as a result of the increase of dairy farm herd sizes in the United States. As a dairyman increases the size of his herd, he must also increase the capacity of his milking parlor in order to harvest the additional milk. This increase in parlor sizes has resulted in tremendous increases in milk throughput and cooling demand. Today's larger farms produce milk at a rate which direct expansion refrigeration systems on bulk milk coolers cannot cool in a timely manner. PHE's are typically utilized in this instance to rapidly cool the milk to the desired temperature (or close to it) before it reaches the bulk milk tank. Typically, ground water is still utilized to provide some initial cooling to bring the milk to between 55 and 70 °F (21 °C). A second (and sometimes third) section of the PHE is added to remove the remaining heat with a mixture of chilled pure water and propylene glycol. These chiller systems can be made to incorporate large evaporator surface areas and high chilled water flow rates to cool high flow rates of milk.

Milking Operation

Milking machines are held in place automatically by a vacuum system that draws the ambient air pressure down from 15 to 21 pounds per square inch (100 to 140 kPa) of vacuum. The vacuum is also used to lift milk vertically through small diameter hoses, into the receiving can. A milk lift pump draws the milk from the receiving can through large diameter stainless steel piping, through the plate cooler, then into a refrigerated bulk tank.

Milk is extracted from the cow's udder by flexible rubber sheaths known as liners or inflations that are surrounded by a rigid air chamber. A pulsating flow of ambient air and vacuum is applied to the inflation's air chamber during the milking process. When ambient air is allowed to enter the chamber, the vacuum inside the inflation causes the inflation to collapse around the cow's teat, squeezing the milk out of teat in a similar fashion as a baby calf's mouth massaging the teat. When the vacuum is reapplied in the chamber the flexible rubber inflation relaxes and opens up, preparing for the next squeezing cycle.

It takes the average cow three to five minutes to give her milk. Some cows are faster or slower. Slow-milking cows may take up to fifteen minutes to let down all their milk. Though milking speed is not related to the quality of milk produced by the cow, it does impact the management of the milking process. Because most milkers milk cattle in groups, the milker can only process a group of cows at the speed of the slowest-milking cow. For this reason, many farmers will group slow-milking cows so as not to stress the faster milking cows.

The extracted milk passes through a strainer and plate heat exchangers before entering the tank, where it can be stored safely for a few days at approximately 40 °F (4 °C). At pre-arranged times, a milk truck arrives and pumps the milk from the tank for transport to a dairy factory where it will be pasteurized and processed into many products. The frequency of pick up depends and the production and storage capacity of the dairy; large dairies will have milk pick-ups once per day.

Management of the Herd

Modern dairy farmers use milking machines and sophisticated plumbing systems to harvest and store the milk from the cows, which are usually milked two or three times daily. In New Zealand, some farmers seeking a better life style, are milking only once per day, trading a slight reduction in production of milk for increased leisure time. During the summer months, cows may be turned out to graze in pastures, both day and night, and are brought into the barn to be milked.

Barns may also incorporate tunnel ventilation into the architecture of the barn structure. This ventilation system is highly efficient and involves opening both ends of the structure allowing cool air to blow through the building. Farmers with this type of structure keep cows inside during the summer months to prevent heat stress, sunburn and damage to udders. During the winter months the cows may be kept in the barn, which is warmed by their collective body heat. Even in winter, the heat produced by the cattle requires the barns to be ventilated for cooling purposes. Many large, modern facilities, and particularly those in tropical areas, keep all animals inside at all times to facilitate herd management.

Farmers typically grow their own food for their cattle. Crops grown may include corn, alfalfa, timothy, wheat, oats, sorghum and clover. These plants are often processed after harvest to preserve or improve nutrient value and prevent spoiling. Corn, alfalfa, wheat, oats, and sorghum crops are often anaerobically fermented to create silage. Many crops such as alfalfa, timothy, oats, and clover are allowed to dry in the field after cutting before being baled into hay.

In the southern hemisphere such as in Australia and New Zealand, cows spend most of their lives outside on pasture, although they may receive supplementation during periods of low pasture availability. Typical supplementary feeds in Australasia are hay, silage or ground maize. The trend in New Zealand is towards feeding cows on a concrete pad to prevent loss of feed by trampling. In New Zealand slower growing winter pasture is rationed. It is carefully controlled by light weight portable electric break feeding fences run on mains power that can be easily repositioned.

Concerns

Animal Waste From Large Cattle Dairies

As measured in phosphorus, the waste output of 5,000 cows roughly equals a municipality of 70,000 people. In the U.S., dairy operations with more than 1,000 cows meet the EPA definition of a CAFO (Concentrated Animal Feeding Operation), and are subject to EPA regulations. For example, in the San Joaquin Valley of California a number of dairies have been established on a very large scale. Each dairy consists of several modern milking parlor set-ups operated as a single enterprise. Each milking parlor is surrounded by a set of 3 or 4 loafing barns housing 1,500 or 2,000 cattle. Some of the larger dairies have planned 10 or more series of loafing barns and milking parlors in this arrangement, so that the total operation may include as many as 15,000 or 20,000 cows. The milking process for these dairies is similar to a smaller dairy with a single milking parlor but repeated several times. The size and concentration of cattle creates major environmental issues associated with manure handling and disposal, which requires substantial areas of cropland (a ratio of 5 or 6 cows to the acre, or several thousand acres for dairies of this size) for manure spreading and dispersion, or several-acre methane digesters. Air pollution from methane gas associated with manure management also is a major concern. As a result, proposals to develop dairies of this size can be controversial and provoke substantial opposition from environmentalists including the Sierra Club and local activists.

Dairy CAFO—EPA

The potential impact of large dairies was demonstrated when a massive manure spill occurred on a 5,000-cow dairy in Upstate New York, contaminating a 20-mile (32 km) stretch of the Black River, and killing 375,000 fish. On 10 August 2005, a manure storage lagoon collapsed releasing 3,000,000 US gallons (11,000,000 l; 2,500,000 imp gal) of manure into the Black River.

Subsequently the New York Department of Environmental Conservation mandated a settlement package of $2.2 million against the dairy.

When properly managed, dairy and other livestock waste, due to its nutrient content (N, P, K), makes an excellent fertilizer promoting crop growth, increasing soil organic matter, and improving overall soil fertility and tilth characteristics. Most dairy farms in the United States are required to develop nutrient management plans for their farms, to help balance the flow of nutrients and reduce the risks of environmental pollution. These plans encourage producers to monitor all nutrients coming onto the farm as feed, forage, animals, fertilizer, etc. and all nutrients exiting the farm as product, crop, animals, manure, etc. For example, a precision approach to animal feeding results in less overfeeding of nutrients and a subsequent decrease in environmental excretion of nutrients, such as phosphorus. In recent years, nutritionists have realized that requirements for phosphorus are much lower than previously thought. These changes have allowed dairy producers to reduce the amount of phosphorus being fed to their cows with a reduction in environmental pollution.

Use of Hormones

It is possible to maintain higher milk production by supplementing cows with growth hormones known as recombinant BST or rBST, but this is controversial due to its effects on animal and possibly human health. The European Union, Japan, Australia, New Zealand and Canada have banned its use due to these concerns.

In the US however, no such prohibition exists, and approximately 17.2% of dairy cows are treated in this way. The U.S. Food and Drug Administration states that no "significant difference" has been found between milk from treated and non-treated cows but based on consumer concerns several milk purchasers and resellers have elected not to purchase milk produced with rBST.

Animal Welfare

The practice of dairy production in a factory farm environment has been criticized by animal welfare activists. Some of the ethical complaints regarding dairy production cited include how often the dairy cattle must remain pregnant, the separation of calves from their mothers, how dairy cattle are housed and environmental concerns regarding dairy production.

The production of milk requires that the cow be in lactation, which is a result of the cow having given birth to a calf. The cycle of insemination, pregnancy, parturition, and lactation, followed by a "dry" period of about two months of forty-five to fifty days, before calving which allows udder tissue to regenerate. A dry period that falls outside this time frame can result in decreased milk production in subsequent lactation.

An important part of the dairy industry is the removal of the calves off the mother's milk after the three days of needed colostrum, allowing for the collection of the milk produced. On some dairies, in order for this to take place, the calves are fed milk replacer, a substitute for the whole milk produced by the cow. Milk replacer is generally a powder, which comes in large bags, and is added to precise amounts of water, and then fed to the calf via bucket or bottle. However, not all dairies use milk replacer - some continue to feed calves milk from the cows in the milking herd. Some dairies even pasteurize extra milk from the main herd to feed calves.

Milk replacers are classified by three categories: protein source, protein/fat (energy) levels, and medication or additives (e.g. vitamins and minerals). Proteins for the milk replacer come from different sources; the more favorable and more expensive all milk protein (e.g. whey protein- a by-product of the cheese industry) and alternative proteins including soy, animal plasma and wheat gluten. The ideal levels for fat and protein in milk replacer are 10-28% and 18-30%, respectively. The higher the energy levels (fat and protein), the less starter feed (feed which is given to young animals) the animal will consume. Weaning can take place when a calf is consuming at least two pounds of starter feed a day and has been on starter for at least three weeks. Milk replacer has climbed in cost US$15–20 a bag in recent years, so early weaning is economically crucial to effective calf management.

Because of the danger of infection to humans, it is important to maintain the health of milk-producing cattle. Common ailments affecting dairy cows include infectious disease (e.g. mastitis, endometritis and digital dermatitis), metabolic disease (e.g. milk fever and ketosis) and injuries caused by their environment (e.g. hoof and hock lesions).

Lameness is commonly considered one of the most significant animal welfare issues for dairy cattle, and is best defined as any abnormality that causes an animal to change its gait. It can be caused by a number of sources, including infections of the hoof tissue (e.g. fungal infections that cause dermatitis) and physical damage causing bruising or lesions (e.g. ulcers or hemorrhage of the hoof). Housing and management features common in modern dairy farms (such as concrete barn floors, limited access to pasture and suboptimal bed-stall design) have been identified as contributing risk factors to infections and injuries. New dairy farms being built now include non-slip flooring and other features designed to minimize risk to cows when moving between pens and to the milking parlor.

Market

Worldwide

Holstein cows on a dairy farm, Comboyne, New South Wales

Dairy farm in Võru Parish, Estonia

There is a great deal of variation in the pattern of dairy production worldwide. Many countries which are large producers consume most of this internally, while others (in particular New Zealand), export a large percentage of their production. Internal consumption is often in the form of liquid milk, while the bulk of international trade is in processed dairy products such as milk powder.

The milking of cows was traditionally a labor-intensive operation and still is in less developed countries. Small farms need several people to milk and care for only a few dozen cows, though for many farms these employees have traditionally been the children of the farm family, giving rise to the term "family farm".

Advances in technology have mostly led to the radical redefinition of "family farms" in industrialized countries such as Australia, New Zealand, and the United States. With farms of hundreds of cows producing large volumes of milk, the larger and more efficient dairy farms are more able to weather severe changes in milk price and operate profitably, while "traditional" very small farms generally do not have the equity or cash flow to do so. The common public perception of large corporate farms supplanting smaller ones is generally a misconception, as many small family farms expand to take advantage of economies of scale, and incorporate the business to limit the legal liabilities of the owners and simplify such things as tax management.

Before large scale mechanization arrived in the 1950s, keeping a dozen milk cows for the sale of milk was profitable. Now most dairies must have more than one hundred cows being milked at a time in order to be profitable, with other cows and heifers waiting to be "freshened" to join the milking herd . In New Zealand the average herd size, for the 2009/2010 season, is 376 cows.

Worldwide, the largest milk producer is India (more than 55% buffalo milk), the largest cow milk exporter is New Zealand, and the largest importer is China. The European Union with its present 28 member countries produced 158,800,000 metric tons (156,300,000 long tons; 175,000,000 short tons) in 2013(96.8% cow milk), the most by any politico-economic union.

World total milk production in 2009 FAO statistics (including cow/buffalo/goat/sheep/camel milk)		
Rank	**Country**	**Production (10^6 kg/y)**
	World	**696,554**
1	India	110,040
2	United States	85,859
3	China	40,553
4	Pakistan	34,362
5	Russia	32,562
6	Germany	28,691
7	Brazil	27,716
8	France	24,218
9	New Zealand	15,217

10	United Kingdom	13,237
11	Italy	12,836
12	Turkey	12,542
13	Poland	12,467
14	Ukraine	11,610
15	Netherlands	11,469
16	Mexico	10,931
17	Argentina	10,500
18	Australia	9,388
19	Canada	8,213
20	Japan	7,909

European Union

Production building at a dairy farm in Norway.

The European Union with its present 27 member countries is the largest milk producer in the world. The largest producers within the EU are Germany and France.

Dairy production in the EU is heavily distorted due to the Common Agricultural Policy – being subsidized in some areas, and subject to production quotas in other.

European total milk production in 2009 FAO statistics (including cow/goat/sheep/buffalo milk)		
Rank	**Country**	**Production (10^6 kg/y)**
	European Union (all 27 countries)	**153,033**
1	Germany	28,691
2	France	24,218
3	United Kingdom	13,237
4	Italy	12,836
5	Poland	12,467

6	Netherlands	11,469
7	Spain	7,252
8	Romania	5,809
9	Ireland	5.373
10	Denmark	4,814

Israel

The dairy farm on Sa'ad was the Israeli leader in 2011 for productivity with an average of 13,785 litres (3,032 imp gal; 3,642 US gal) per head that year. A dairy cow named Kharta, was the world record holder giving 18,208 litres (4,005 imp gal; 4,810 US gal) liters of milk. The 954 Israeli dairy farms achieved a world leading average production of 11,775 litres (2,590 imp gal; 3,111 US gal) a year per head, while the national average per head was 10,336 litres (2,274 imp gal; 2,730 US gal). Israeli consumption is lower than other western countries with an average of 180 litres (40 imp gal; 48 US gal) per person.

United States

In the United States, the top five dairy states are, in order by total milk production; California, Wisconsin, New York, Idaho, and Pennsylvania. Dairy farming is also an important industry in Florida, Minnesota, Ohio and Vermont. There are 65,000 dairy farms in the United States.

Pennsylvania has 8,500 farms with 555,000 dairy cows. Milk produced in Pennsylvania yields an annual revenue of about US$1.5 billion.

Milk prices collapsed in 2009. Senator Bernie Sanders accused Dean Foods of controlling 40% of the country's milk market. He has requested the United States Department of Justice to pursue an anti-trust investigation. Dean Foods says it buys 15% of the country's raw milk. In 2011, a federal judge approved a settlement of $30 million to 9,000 farmers in the Northeast.

Herd size in the US varies between 1,200 on the West Coast and Southwest, where large farms are commonplace, to roughly 50 in the Midwest and Northeast, where land-base is a significant limiting factor to herd size. The average herd size in the U.S. is about one hundred cows per farm but the median size is 900 cows with 49% of all cows residing on farms of 1000 or more cows.

Poultry Farming

Poultry farming is the raising of domesticated birds such as chickens, ducks, turkeys and geese for the purpose of farming meat or eggs for food. Poultry are farmed in great numbers with chickens being the most numerous. More than 50 billion chickens are raised annually as a source of food, for both their meat and their eggs. Chickens raised for eggs are usually called layers while chickens raised for meat are often called broilers. In the US, the national organization overseeing poultry production is the Food and Drug Administration (FDA). In the UK, the national organisation is the Department for Environment, Food and Rural Affairs (Defra).

Intensive and Alternative

According to the researchers and scientists, 74% of the world's poultry meat, and 68 percent of eggs are produced in ways that are described as 'intensive'. One alternative to intensive poultry farming is free-range farming using lower stocking densities. Poultry producers routinely use nationally approved medications, such as antibiotics, in feed or drinking water, to treat disease or to prevent disease outbreaks. Some FDA-approved medications are also approved for improved feed utilization.

Egg-laying Chickens – Husbandry Systems

Commercial hens usually begin laying eggs at 16–20 weeks of age, although production gradually declines soon after from approximately 25 weeks of age. This means that in many countries, by approximately 72 weeks of age, flocks are considered economically unviable and are slaughtered after approximately 12 months of egg production, although chickens will naturally live for 6 or more years. In some countries, hens are force moulted to re-invigorate egg-laying.

Environmental conditions are often automatically controlled in egg-laying systems. For example, the duration of the light phase is initially increased to prompt the beginning of egg-laying at 16–20 weeks of age and then mimics summer day length which stimulates the hens to continue laying eggs all year round; normally, egg production occurs only in the warmer months. Some commercial breeds of hen can produce over 300 eggs a year!

Free-range

Commercial free range hens

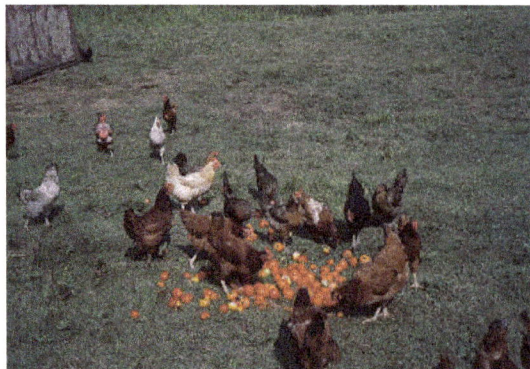

Free range chickens being fed outdoors

Free-range poultry farming allows chickens to roam freely for a period of the day, although they are usually confined in sheds at night to protect them from predators or kept indoors if the weather is particularly bad. In the UK, the Department for Environment, Food and Rural Affairs (Defra) states that a free-range chicken must have day-time access to open-air runs during at least half of its life. Unlike in the United States, this definition also applies to free-range egg laying hens. The European Union regulates marketing standards for egg farming which specifies a minimum condition for free-range eggs that "hens have continuous daytime access to open-air runs, except in the case of temporary restrictions imposed by veterinary authorities". The RSPCA "Welfare standards for laying hens and pullets" indicates that the stocking rate must not exceed 1,000 birds per hectare (10 m² per hen) of range available and a minimum area of overhead shade/shelter of 8 m² per 1,000 hens must be provided.

Free-range farming of egg-laying hens is increasing its share of the market. Defra figures indicate that 45% of eggs produced in the UK throughout 2010 were free-range, 5% were produced in barn systems and 50% from cages. This compares with 41% being free-range in 2009.

Suitable land requires adequate drainage to minimise worms and coccidial oocysts, suitable protection from prevailing winds, good ventilation, access and protection from predators. Excess heat, cold or damp can have a harmful effect on the animals and their productivity. Free-range farmers have less control than farmers using cages in what food their chickens eat, which can lead to unreliable productivity, though supplementary feeding reduces this uncertainty. In some farms, the manure from free-range poultry can be used to benefit crops.

The benefits of free-range poultry farming for laying hens include opportunities for natural behaviours such as pecking, scratching, foraging and exercise outdoors.

Both intensive and free-range farming have animal welfare concerns. Cannibalism, feather pecking and vent pecking can be common, prompting some farmers to use beak trimming as a preventative measure, although reducing stocking rates would eliminate these problems. Diseases can be common and the animals are vulnerable to predators. Barn systems have been found to have the worst bird welfare. In South-East Asia, a lack of disease control in free range farming has been associated with outbreaks of Avian influenza.

Organic

In organic egg-laying systems, chickens are also free-range. Organic systems are based upon restrictions on the routine use of synthetic yolk colourants, in-feed or in-water medications, other food additives and synthetic amino acids, and a lower stocking density and smaller group sizes. The Soil Association standards used to certify organic flocks in the UK, indicate a maximum outdoors stocking density of 1,000 birds per hectare and a maximum of 2,000 hens in each poultry house. In the UK, organic laying hens are not routinely beak-trimmed.

Yarding

While often confused with free-range farming, yarding is actually a separate method of poultry culture by which chickens and cows are raised together. The distinction is that free-range poultry are either totally unfenced, or the fence is so distant that it has little influence on their freedom of movement. Yarding is common technique used by small farms in the Northeastern US. The birds are released daily from hutches or coops. The hens usually lay eggs either on the floor of the coop or in baskets if provided by the farmer. This husbandry technique can be complicated if used with roosters, mostly because of aggressive behavior.

Battery Cage

Battery Cages for Layer Hens

Bank of cages

The majority of hens in many countries are reared in battery cages, although the European Union Council Directive 1999/74/EC has banned the conventional battery cage in EU states from January 2012. These are small cages, usually made of metal in modern systems, housing 3 to 8 hens. The walls are made of either solid metal or mesh, and the floor is sloped wire mesh to allow the faeces to drop through and eggs to roll onto an egg-collecting conveyor belt. Water is usually provided by overhead nipple systems, and food in a trough along the front of the cage replenished at regular intervals by a mechanical chain.

The cages are arranged in long rows as multiple tiers, often with cages back-to-back (hence the term 'battery cage'). Within a single shed, there may be several floors containing battery cages meaning that a single shed may contain many tens of thousands of hens. Light intensity is often kept low (e.g. 10 lux) to reduce feather pecking and vent pecking. Benefits of battery cages include easier care for the birds, floor eggs which are expensive to collect are eliminated, eggs are cleaner, capture at the end of lay is expedited, generally less feed is required to produce eggs, broodiness is eliminated, more hens may be housed in a given house floor space, internal parasites are more easily treated, and labor requirements are generally much reduced.

In farms using cages for egg production, there are more birds per unit area; this allows for greater productivity and lower food costs. Floor space ranges upwards from 300 cm^2 per hen. EU standards in 2003 called for at least 550 cm^2 per hen. In the US, the current recommendation by the United

Egg Producers is 67 to 86 in² (430 to 560 cm²) per bird. The space available to battery hens has often been described as less than the size of a piece of A4 paper. Animal welfare scientists have been critical of battery cages because they do not provide hens with sufficient space to stand, walk, flap their wings, perch, or make a nest, and it is widely considered that hens suffer through boredom and frustration through being unable to perform these behaviours. This can lead to a wide range of abnormal behaviours, some of which are injurious to the hens or their cagemates.

Furnished Cage

In 1999, the European Union Council Directive 1999/74/EC banned conventional battery cages for laying hens throughout the European Union from January 1, 2012; they were banned previously in other countries including Switzerland. In response to these bans, development of prototype commercial furnished cage systems began in the 1980s. Furnished cages, sometimes called 'enriched' or 'modified' cages, are cages for egg laying hens which have been designed to overcome some of the welfare concerns of battery cages whilst retaining their economic and husbandry advantages, and also provide some of the welfare advantages of non-cage systems. Many design features of furnished cages have been incorporated because research in animal welfare science has shown them to be of benefit to the hens. In the UK, the Defra "Code for the Welfare of Laying Hens" states furnished cages should provide at least 750 cm² of cage area per hen, 600 cm² of which should be usable; the height of the cage other than that above the usable area should be at least 20 cm at every point and no cage should have a total area that is less than 2000 cm². In addition, furnished cages should provide a nest, litter such that pecking and scratching are possible, appropriate perches allowing at least 15 cm per hen, a claw-shortening device, and a feed trough which may be used without restriction providing 12 cm per hen.

Modern egg laying breeds often suffer from osteoporosis which results in the chicken's skeletal system being weakened. During egg production, large amounts of calcium are transferred from bones to create egg-shell. Although dietary calcium levels are adequate, absorption of dietary calcium is not always sufficient, given the intensity of production, to fully replenish bone calcium. This can lead to increases in bone breakages, particularly when the hens are being removed from cages at the end of laying.

Countries such as Austria, Belgium or Germany are planning to ban furnished cages until 2025 additionally to the already banned conventional cages.

Meat-Producing Chickens – Husbandry Systems

Broilers in a production house

Indoor Broilers

Meat chickens, commonly called broilers, are floor-raised on litter such as wood shavings, peanut shells, and rice hulls, indoors in climate-controlled housing. Under modern farming methods, meat chickens reared indoors reach slaughter weight at 5 to 9 weeks of age. The first week of chickens life they can grow 300 percent of their body size, a nine-week-old chicken can average over 9 pounds in body weight. At nine weeks a hen will average around 7 pounds and a rooster will weigh around 12 pounds, having a nine-pound average.

Broilers are not raised in cages. They are raised in large, open structures known as grow out houses. A farmer receives the birds from the hatchery at one day old. A grow out consist of 5 to 9 weeks according on how big the kill plant wants the chickens to be. These houses are equipped with mechanical systems to deliver feed and water to the birds. They have ventilation systems and heaters that function as needed. The floor of the house is covered with bedding material consisting of wood chips, rice hulls, or peanut shells. In some cases they can be grown over dry litter or compost. Because dry bedding helps maintain flock health, most growout houses have enclosed watering systems ("nipple drinkers") which reduce spillage.

Keeping birds inside a house protects them from predators such as hawks and foxes. Some houses are equipped with curtain walls, which can be rolled up in good weather to admit natural light and fresh air. Most growout houses built in recent years feature "tunnel ventilation," in which a bank of fans draws fresh air through the house.

Traditionally, a flock of broilers consist of about 20,000 birds in a growout house that measures 400/500 feet long and 40/50 feet wide, thus providing about eight-tenths of a square foot per bird. The Council for Agricultural Science and Technology (CAST) states that the minimum space is one-half square foot per bird. More modern houses are often larger and contain more birds, but the floor space allotment still meets the needs of the birds. The larger the bird is grown the fewer chickens are put in each house, to give the bigger bird more space per square foot.

Because broilers are relatively young and have not reached sexual maturity, they exhibit very little aggressive conduct.

Chicken feed consists primarily of corn and soybean meal with the addition of essential vitamins and minerals. No hormones or steroids are allowed in raising chickens.

Issues with Indoor Husbandry

In intensive broiler sheds, the air can become highly polluted with ammonia from the droppings. In this case a farmer must run more fans to bring in more clean fresh air. If not this can damage the chickens' eyes and respiratory systems and can cause painful burns on their legs (called hock burns) and blisters on their feet. Broilers bred for fast growth have a high rate of leg deformities because the large breast muscles causes distortions of the developing legs and pelvis, and the birds cannot support their increased body weight. In cases where the chickens become crippled and can't walk farmers have to go in and pull them out. Because they cannot move easily, the chickens are not able to adjust their environment to avoid heat, cold or dirt as they would in natural conditions. The added weight and overcrowding also puts a strain on their hearts and lungs and Ascites can develop. In the UK, up to 19 million broilers die in their sheds from heart failure each year. In the case of no ventilation due to power failure during a heat wave 20,000 chicken can die in a short period of time. In a good grow out a farmer should sell between 92 and 96 percent of their flock. With a 1.80 to a 2.0 feed conversion ratio. After the marking of birds the farmer must clean out and repair for another flock. A farmer should average 4 to 5 grow outs a year.

Indoor with Higher Welfare

Chickens are kept indoors but with more space (around 12 to 14 birds per square metre). They have a richer environment for example with natural light or straw bales that encourage foraging and perching. The chickens grow more slowly and live for up to two weeks longer than intensively farmed birds. The benefits of higher welfare indoor systems are the reduced growth rate, less crowding and more opportunities for natural behaviour.

Free-Range Broilers

Turkeys on pasture at an organic farm

Free-range broilers are reared under similar conditions to free-range egg laying hens. The breeds grow more slowly than those used for indoor rearing and usually reach slaughter weight at approximately 8 weeks of age. In the EU, each chicken must have one square metre of outdoor space. The benefits of free-range poultry farming include opportunities for natural behaviours such as pecking, scratching, foraging and exercise outdoors. Because they grow slower and have opportunities for exercise, free-range broilers often have better leg and heart health.

Organic Broilers

Organic broiler chickens are reared under similar conditions to free-range broilers but with restrictions on the routine use of in-feed or in-water medications, other food additives and synthetic amino acids. The breeds used are slower growing, more traditional breeds and typically reach slaughter weight at around 12 weeks of age. They have a larger space allowance outside (at least 2 square metres and sometimes up to 10 square metres per bird). The Soil Association standards indicate a maximum outdoors stocking density of 2,500 birds per hectare and a maximum of 1,000 broilers per poultry house.

Issues

Humane Treatment

Battery cages

Chickens transported in a truck.

Animal welfare groups have frequently criticized the poultry industry for engaging in practices which they believe to be inhumane. Many animal rights advocates object to killing chickens for food, the "factory farm conditions" under which they are raised, methods of transport, and slaughter. Compassion Over Killing and other groups have repeatedly conducted undercover investigations at chicken farms and slaughterhouses which they allege confirm their claims of cruelty.

Conditions in chicken farms may be unsanitary, allowing the proliferation of diseases such as salmonella, E. coli and campylobacter. Chickens may be raised in very low light intensities, sometimes total darkness, to reduce injurious pecking. Concerns have been raised that companies growing single varieties of birds for eggs or meat are increasing their susceptibility to disease. Rough handling, crowded transport during various weather conditions and the failure of existing stunning systems to render the birds unconscious before slaughter, have also been cited as welfare concerns.

A common practice among hatcheries for egg-laying hens is the culling of newly hatched male chicks since they do not lay eggs and do not grow fast enough to be profitable for meat. There are plans to more ethically destroy the eggs before the chicks are hatched by "in-ovo" sex determination.

Beak Trimming

Laying hens are routinely beak-trimmed at 1 day of age to reduce the damaging effects of aggression, feather pecking and cannibalism. Scientific studies (see below) have shown that beak trimming is likely to cause both acute and chronic pain.

The beak is a complex, functional organ with an extensive nervous supply including nociceptors that sense pain and noxious stimuli. These would almost certainly be stimulated during beak trimming, indicating strongly that acute pain would be experienced. Behavioural evidence of pain after beak trimming in layer hen chicks has been based on the observed reduction in pecking behavior, reduced activity and social behavior, and increased sleep duration. Severe beak trimming, or beak trimming birds at an older age, may cause chronic pain. Following beak trimming of older or adult hens, the nociceptors in the beak stump show abnormal patterns of neural discharge, which indicate acute pain.

Neuromas, tangled masses of swollen regenerating axon sprouts, are found in the healed stumps of birds beak trimmed at 5 weeks of age or older and in severely beak trimmed birds. Neuromas have been associated with phantom pain in human amputees and have therefore been linked to chronic pain in beak trimmed birds. If beak trimming is severe because of improper procedure or done in older birds, the neuromas will persist which suggests that beak trimmed older birds experience chronic pain, although this has been debated.

Beak-trimmed chicks will initially peck less than non-trimmed chickens, which animal behavioralist Temple Grandin attributes to guarding against pain. The animal rights activist, Peter Singer, claims this procedure is bad because beaks are sensitive, and the usual practice of trimming them without anaesthesia is considered inhumane by some. Some within the chicken industry claim that beak-trimming is not painful whereas others argue that the procedure causes chronic pain and discomfort, and decreases the ability to eat or drink.

Antibiotics

Overview of Antibiotic use in Poultry

Antibiotics have been used in poultry farming in mass quantities since 1951, when the Food and Drug Administration (FDA) approved their use. Three years prior to the FDA's approval, scientists were investigating a phenomena in which chickens who were rooting through bacteria-rich manure were displaying signs of greater health than those who did not. Through testing, it was discovered that chickens who were fed a variety of vitamin B12 manufactured with the residue of a certain antibiotic grew 50 percent faster than those chickens who were fed B12 manufactured from a different source. Further testing confirmed that use of antibiotics did improve the health of the chickens, resulting in the chickens laying more eggs and experiencing lower mortality rates and less illness. Upon this discovery, farmers transitioned from expensive animal proteins to comparatively inexpensive antibiotics and B12. Chickens were now reaching their market weight at a much faster rate and at a lower cost. With a growing population and greater demand on the farmers, antibiotics appeared to be an ideal and cost-effective way to increase the output of poultry. Since this discovery, antibiotics have been routinely used in poultry production, but more recently have been the topic of debate secondary to the fear of bacterial antibiotic resistance.

Emerging Threats: Antibiotic Resistance

The Centers for Disease Control (CDC), has identified the emergence of antibiotic resistance as a national threat. The concern over antibiotic use in livestock arises from the necessity antibiotics have in keeping populations disease-free. As of 2016, over 70 percent of FDA approved antibiotics are utilized in modern, high production poultry farms to prevent, control, and treat disease. The FDA released a report in 2009 estimating that 29 million pounds of antibiotics had been used in livestock in that year alone. However, surveillance of consumer exposure to antibiotics through poultry consumption is limited. More specifically in 2012, the FDA speculated the most significant public health threat in regard to antimicrobial use in animals is the exposure of antimicrobial resistant bacteria to humans. These statements are challenged by the American meat industry lobbyists that antibiotics are used responsibly and judiciously in order to ensure effectiveness.

Consumer Health effects

Consumers are exposed to antibiotic resistance through consumption of poultry products that have prior exposure to resistant strains. In poultry husbandry, the practice of using medically important antibiotics can select for resistant strains of bacteria, which are then transferred to consumers through poultry meat and eggs. The CDC acknowledges this transferal pathway in their 2013 report of Antibiotic Resistant Threats in the United States. The annual rate of foodborne illness in the United States is one in six. For the 48 million individuals affected, antibiotics play a critical role in thwarting mortality rates. In a literature review conducted by the Review of Antimicrobial Resistance 100 out of 139 studies found evidence of a link between antibiotic use in animals and antibiotic resistance in consumers.

When a gram-negative bacterial infection is suspected in a patient, one of the first-line options for treatment is in the fluoroquinolone family. This, along with penicillin, is one of the first families of antibiotics utilized in the broiler industry. If this first-line treatment is not successful, a stronger class of antibiotics is typically used, however, there is a limitation on how many classes are available, as well as which medications are available on hospital formularies. There is also more drug toxicity affiliated with second and third line antibiotic options. This is one example why it is critical to keep as many first line antibiotic options available for human use.

Other issues are associated with duration and complexity of infection. On average, treatment for non-resistant bacteria is administered 11.5 hours after diagnosis, and treatment for resistant bacteria is administered 72 hours after diagnosis. This is a reflection of the additional threat of prolonged incubation, leading to greater potential for systemic disease, with higher morbidity and mortality associated with opportunities for complications, and prolonged treatment time. For example, of the two million people affected by resistant infections a year, 23,000 will die. Severity in mortality is coupled when exposed to high risk populations such as immunocompromised and elderly individuals in hospital and nursing home settings.

History of US Federal Policy on Antibiotic use in Livestock

- 1940s – Beginning of utilization of antibiotics in livestock feed

- 1951 - Antibiotics first FDA approved for use in poultry. Approved uses included production (growth enhancement), treatment, control, or prevention of animal disease. Antibiotics were also available for purchase over the counter at that time.

- 1970 - FDA task force publication proposes limitations of utilizing antibiotics in livestock feed that are also used in humans.

- 1975 - Secondary to this publication, drug sponsors are required to submit studies demonstrating the antibiotics did not harm human health

- 1976 - Stuart Levy study demonstrating tetracycline resistant E. coli moving to consumers

- 1977 - FDA proposal to remove penicillin and tetracycline in subtherapeutic doses, however, request by Congress for further studies to be conducted.

- 1980 - National Academy of Science recruited by the FDA to conduct further studies, specifically for penicillins and tetracyclines. Conclusion from these studies indicated no sufficient evidence to ban these antibiotics.

- 1980s-early 2000s - Further studies continued, supported by the FDA

- 2003 - FDA issued guidance to pharmaceuticals for an approval process utilizing new antibiotics in animal feed. For antibiotics already in use, the FDA would have to withdraw approval for each individual medication.

- 2005 - Enrofloxacin, an already utilized antibiotic, was removed from poultry production. This took 5 years to accomplish.

- 2010 - FDA first draft of "voluntary" limitations of medically important antibiotics in livestock, and requirement of veterinarian oversight, which would later become "Guidance for Industry #209."

- 2011 - FDA removed original request from 1977 to remove penicillins and tetracyclines in feed.

- 2012 - FDA finalized "Guidance for Industry #209," which was implemented under the Veterinary Feed Directives. These guidelines were issued to pharmaceuticals.

- 2013 - FDA issues "Guidance for Industry #213," which provided additional information to pharmaceuticals for recommendations from #209.

- 2014 - All 26 pharmaceutical companies producing antibiotics used in livestock feed agreed to the FDA guidelines in #213. Gave total of 3 years to make all recommended changes.

Current Federal Regulators

National Antimicrobial Resistance Monitoring System's (NARMS) Enteric Bacteria program - Established in 1996, and represents a collaboration between the USDA, FDA, and CDC. Its purpose is to organize these organizations into a drug monitoring program for antibiotics utilized in animal feed withe goal of maintaining their medical efficacy. There are three branches which oversee humans, retail meat, and food animals.

- USDA - Operating under the Food Safety and Inspection Service (FSIS). Main role is in charge of testing imported and domestic meat for antimicrobial resistant bacteria. If a 'residue violation' found, they may condemn the product. Regardless, funding and resources are not available for outbreak investigations at farms or ranches.

- FDA - Operating under the Center for Veterinary Medicine (CVM). Works with CDC to monitor retail meat.

- CDC - Monitors human samples.

Vertical Integration

This is the current business structure utilized almost universally in the broiler, or chicken bred for meat, industry. This also began in the 1940s when antibiotics began to be utilized in livestock feed. Perdue is credited as the pioneer of this structure. The basis is centralization of production. 'Integrators' control cost, policy, and are the decision makers of production. They decide feed formulations, choice of antibiotic administration, and cover those costs in addition to veterinary services. They also own the poultry that is grown. Farmers are labeled as 'Growers' or 'Operators'. They own the land and buildings where the poultry is grown, and are essentially caretakers for the poultry growth to the Integrators. The benefit for Growers in this business structure is they are guaranteed payment from the Integrators, which is compensated in weight gained by each flock. Due to this structure, about 90% of broilers are raised within 60 miles of the processing plant. Integrators are large poultry companies such as Perdue, Tyson, Pilgrim's Pride, Koch Foods, etc. There are about 20 of these companies in the U.S. that control 96% of all broilers produced in 2011.

Regulatory Surveys

There are two main surveys distributed to farmers by the federal government to aid in various regulations of the agricultural industry. They are the Agricultural and Resource Management Survey (ARMS) and the National Animal Health Monitoring Survey (NAHMS).

Agricultural and Resource Management Survey (ARMS) - Ran by the USDA's Economic Research Service (ERS) and National Agricultural Statistics Service (NASS). The main focus is finances of farming, production practices, and resource use. Seventeen total states are sampled every 5–6 years per livestock type, with the most recent surveys distributed to broiler farmers in 2006 and 2011. There was one question about utilization of antibiotics in poultry food or water, excluding use for illness treatment.

Antibiotic Resistant Outbreaks from Poultry Meat

In order to minimize and prevent any residues of antibiotics in chicken meat, any chickens given antibiotics are required to have a "withdrawal" period before they can be slaughtered. Samples of poultry at slaughter are randomly tested by the FSIS, and show a very low percentage of residue violations. Although violations are minimal, these small amounts of antibiotics have still contributed to antibiotic resistant outbreaks in the U.S. There are five infectious agents that account for 90% of foodborne related deaths. Three consistently found in poultry are: Salmonella, Campylobacter, and Escherichia coli.

- 2014: Outbreak of Salmonella in 634 people across 29 states (38% hospitalized) from eating chicken from Foster Farms that was sold at Costco. 44/68 tested isolates were resistant to at least 1 drug (65%), and 4 of 5 chicken samples tested were drug resistant (80%).

- 2015: Outbreak of Salmonella in 15 people in 7 states (4 hospitalized) from eating frozen stuffed chicken produced by Barber Foods.

Limitations & Challenges

One obstacle to gathering more comprehensive data on the use of antibiotics in feed is the majority of the poultry industry utilizes vertical integration. As a consequence, farmers are often unaware of what components go into the feed, including whether or not antibiotics are used. Also in antibiotic usage in general, there are criteria to define bacterial resistance to specific antibiotics, however, there are no standards to divide the bacteria into resistant and susceptible categories based on antibiotics utilized.

The poultry industry also plays a large part in the United States economy, both in domestic purchasing and through international demand. The USDA reports that the U.S. is the "world's largest producer and second largest exporter of poultry meat." In 2010, the U.S. produced 36.9 billion pounds of broiler meat and exported 6.8 billion pounds of broiler meat. This equates to an estimated retail value of 45 billion dollars in 2010.

Both the agricultural and pharmaceutical industries have been lobbying against legislation that seeks to quell non-therapeutic antibiotic use in livestock since the first introduction of such legislation in Congress in the 1970's. Despite scientific evidence suggesting a strong association between antibiotic use in poultry and other livestock, agribusiness lobbies such as The National Chicken Council argue that there is not sufficient evidence to purport that there is a measurable impact to humans and shifts the blame of the problem of antibiotic resistance to overprescribing in the field of medicine.

With antibiotic restrictions, integrators will bare the immediate costs of these changes, and would likely result in modified finances and contracts with growers. Also, public health agencies may not have adequate scientific evidence for making appropriate decisions for better public health outcomes, secondary to lack of research funds. As a reference, the US spends about $101 billion per year for both governmental and biomedical industrial research, which is only 5% of total health expenditures.

Solutions

Several policies have been proposed to improve data collection and transparency in livestock production. For example, the 2013 Delivering Antimicrobial Transparency in Animals (DATA) Act proposed the enactment of policies to acquire more accurate documentation of antibiotic use in growth promotion by farmers, drug manufacturers, and the FDA. Also, the Preservation of Antibiotics for Medical Treatment Act (PAMTA) was enacted to eliminate the use of medically important antibiotics in livestock. In 2015, the Preventing Antibiotic Resistance Act (PARA) was passed with two components: requirement of drug companies to provide evidence that antibiotics that are approved for use in poultry, and that meat production does not add to the growing threat of antibiotic resistance in humans. Antimicrobial Stewardship Programs (ASPs) serve as an example of systematic monitoring and analysis of data via interdisciplinary and multi-sectoral collaboration.

Performing quality improvement in the process of livestock production is another focus. Some alternative methods include "improving hygiene, using enzymes, probiotics, prebiotics, and acids to improve health and utilizing bacteriocins, antimicrobial peptides, and bacteriophages as substitutes for antibiotics." Adaptations of methods by other countries is an addtional focus. For example, the use of antibiotics in feed was banned in Sweden in 1985 with no compensatory increase in antibiotic usage in other sectors of production, proving that a ban can be successfully administered without unintended impacts on other categories.

Major producers in the poultry industry have also begun to make strides towards change, largely due to public concern over the widespread use of antibiotics in poultry. Some producers have started eliminating the use of antibiotics in order to produce and market chickens that may legally be labeled "antibiotic free". In 2007, Perdue began phasing out all medically important antibiotics from its feed and hatcheries and began selling poultry products labeled "no antibiotics ever" under the Harvestland brand. Consumer response was positive and in 2014 Perdue also began phasing out ionophores from its hatchery and began using the "antibiotic free" labels on its Harvestland, Simply Smart and Perfect Portions products.

Impacts of Change

As Guidance for Industry #213 has been voluntarily accepted, it will be a violation of the Federal Food, Drug, and Cosmetic Act to use antibiotics in livestock production for non-therapeutic purposes. However, as there is now a requirement for veterinary oversight and approval for antibiotics use, there is leeway in the interpretation of non-therapeutic purposes dependent on the situation. For example, per the FDA, "a veterinarian may determine, based on the client's production practices and history, that weaned beef calves arriving at a feedlot in bad weather after a lengthy transport are at risk to develop bacterial respiratory infection. In this case, the veterinarian might choose to preventively treat these calves with an antimicrobial approved for prevention of that bacterial infection."

The FDA is not trying to regulate all antimicrobials at this time - only those antibiotics which are considered "medically important." For example, bacitracin, a common antibiotic found in over the counter antibiotic ointments, is not classified as "medically important." Also, ionophores, which are not apart of human medicine but given for improving the health of livestock, are also not included in this regulation.

Arsenic

Poultry feed can also include roxarsone or nitarsone, arsenical antimicrobial drugs that also promote growth. Roxarsone was used as a broiler starter by about 70% of the broiler growers between 1995 and 2000. The drugs have generated controversy because it contains arsenic, which is highly toxic to humans. This arsenic could be transmitted through run-off from the poultry yards. A 2004 study by the U.S. magazine Consumer Reports reported "no detectable arsenic in our samples of muscle" but found "A few of our chicken-liver samples has an amount that according to EPA standards could cause neurological problems in a child who ate 2 ounces of cooked liver per week or in an adult who ate 5.5 ounces per week." The U.S. Food and Drug Administration (FDA), however, is the organization responsible for the regulation of foods in America, and all samples tested were "far less than the... amount allowed in a food product."

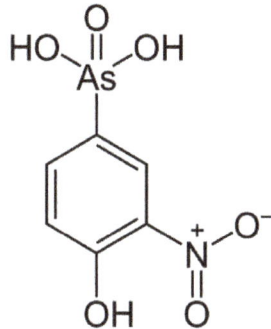

Roxarsone, a controversial arsenic compound used as a nutritional supplement for chickens.

Growth Hormones

Hormone use in poultry production is illegal in the United States. Similarly, no chicken meat for sale in Australia is fed hormones. Several scientific studies have documented the fact that chickens grow rapidly because they are bred to do so, not because of growth hormones. A small producer of natural and organic chickens confirmed this assumption:

E. coli

According to Consumer Reports, "1.1 million or more Americans [are] sickened each year by undercooked, tainted chicken." A USDA study discovered *E. coli (Biotype I)* in 99% of supermarket chicken, the result of chicken butchering not being a sterile process. However, the same study also shows that the strain of *E. coli* found was always a non-lethal form, and no chicken had any of the pathenogenic O157:H7 serotype. Many of these chickens, furthermore, had relatively low levels of contamination.

Feces tend to leak from the carcass until the evisceration stage, and the evisceration stage itself gives an opportunity for the interior of the carcass to receive intestinal bacteria. (The skin of the carcass does as well, but the skin presents a better barrier to bacteria and reaches higher temperatures during cooking.) Before 1950, this was contained largely by not eviscerating the carcass at the time of butchering, deferring this until the time of retail sale or in the home. This gave the intestinal bacteria less opportunity to colonize the edible meat. The development of the "ready-to-cook broiler" in the 1950s added convenience while introducing risk, under the assumption that end-to-end refrigeration and thorough cooking would provide adequate protection. *E. coli* can be killed by proper cooking times, but there is still some risk associated with it, and its near-ubiquity in commercially farmed chicken is troubling to some. Irradiation has been proposed as a means of sterilizing chicken meat after butchering.

The aerobic bacteria found in poultry housing can include not only *E. coli*, but *Staphylococcus*, *Pseudomona*, *Micrococcus* and others as well. These contaminants can contribute to dust that often cause issues with the respiratory systems of both the poultry and humans working in the environment. If bacterial levels in the poultry drinking water reach high levels, it can result in bacterial diarrhoea which can lead to blood poisoning should the bacteria spread from the damaged intestines.

Salmonella too can be stressful on poultry production. How it causes disease has been investigated in some detail.

Avian Influenza

There is also a risk that crowded conditions in chicken farms will allow avian influenza (bird flu) to spread quickly. A United Nations press release states: "Governments, local authorities and international agencies need to take a greatly increased role in combating the role of factory-farming, commerce in live poultry, and wildlife markets which provide ideal conditions for the virus to spread and mutate into a more dangerous form..."

Efficiency

Farming of chickens on an industrial scale relies largely on high protein feeds derived from soybeans; in the European Union the soybean dominates the protein supply for animal feed, and the poultry industry is the largest consumer of such feed. Two kilograms of grain must be fed to poultry to produce 1 kg of weight gain, much less than that required for pork or beef. However, for every gram of protein consumed, chickens yield only 0.33 g of edible protein.

Economic Factors

Changes in commodity prices for poultry feed have a direct effect on the cost of doing business in the poultry industry. For instance, a significant rise in the price of corn in the United States can put significant economic pressure on large industrial chicken farming operations.

Worker Health and Safety

Poultry workers experience substantially higher rates of illness and injury than manufacturing workers do on average.

Muscular Disorders

For the year 2013, there were an estimated 1.59 cases of occupation-related illness per 100 full time U.S. meat and poultry workers, compared to .36 for manufacturing workers overall. Injuries are associated with repetitive movements, awkward postures, and cold temperatures. High rates of carpal tunnel syndrome and other muscular and skeletal disorders are reported. Disinfectant chemicals and infectious bacteria are causes of respiratory illnesses, allergic reactions, diarrhea, and skin infections.

Respiratory Consequences

Poultry housing has been shown to have adverse effects on the respiratory health of workers, ranging from a cough to chronic bronchitis. Workers are exposed to concentrated airborne particulate matter (PM) and endotoxins (a harmful waste product of bacteria. In a conventional hen house a conveyor belt beneath the cages removes the manure. In a cage-free aviary system the manure coats the ground, resulting in the build-up of dust and bacteria over time. Eggs are often laid on the ground or under cages in the aviary housing, causing workers to come close to the floor and force dust and bacteria into the air, which they then inhale during egg collection.

Excretory Consequences

Oxfam America reports that huge industrialized poultry operations are under such pressure to maximize profits that workers are denied access to restrooms.

World Chicken Population

The Food and Agriculture Organization of the United Nations estimated that in 2002 there were nearly sixteen billion chickens in the world, counting a total population of 15,853,900,000. The figures from the *Global Livestock Production and Health Atlas* for 2004 were as follows:

1. China (3,860,000,000)
2. United States (1,970,000,000)
3. Indonesia (1,200,000,000)
4. Brazil (1,100,000,000)
5. Pakistan (691,948,000)
6. India (648,830,000)
7. Mexico (540,000,000)
8. Russia (340,000,000)
9. Japan (286,000,000)
10. Iran (280,000,000)
11. Turkey (250,000,000)
12. Bangladesh (172,630,000)
13. Nigeria (143,500,000)

In 2009 the annual number of chicken raised was estimated at 50 billion, with 6 billion raised in the European Union, over 9 billion raised in the United States and more than 7 billion in China.

In 1950, the average America consumed 20 pounds of chicken per year, but it is predicted that the average consumption will be 89 pounds in 2015. Additionally, in 1980 most chickens were sold whole, and by 2000 almost 90 percent of chickens were sold after being processed into parts. This increase in consumption and processing has led to many of these occupation-related illness.

Pig Farming

Interior of pig farm at Bjärka-Säby Castle, Sweden, 1911.

Pig farming is the raising and breeding of domestic pigs. It is a branch of animal husbandry. Pigs are raised principally as food (e.g. pork, bacon, gammon) and sometimes for their skin.

Pigs on a farm

A sow suckling her piglets.

Pigs are amenable to many different styles of farming. Intensive commercial units, commercial free range enterprises, extensive farming - being allowed to wander around a village, town or city, or tethered in a simple shelter or kept in a pen outside the owners house. Historically pigs were kept in small numbers and were closely associated with the residence of the owner, or in the same village or town. They were valued as a source of meat, fat and for the ability to turn inedible food into meat, and often fed household food waste if kept on a homestead. Pigs have been farmed to dispose of municipal garbage on a large scale.

All these forms of pig farm are in use today. In developed nations, commercial farms house thousands of pigs in climate-controlled buildings. Pigs are a popular form of livestock, with more than one billion pigs killed each year worldwide, 100 million of them in the USA. The majority of pigs are used for human food but also supply skin, fat and other materials for use as clothing, ingredients for processed foods, cosmetics and other and medical use.

The activities on a pig farm depend on the husbandry style of the farmer, and range from very little intervention (as when pigs are allowed to roam villages or towns and dispose of garbage) to intensive systems where the pigs are contained in a building for the majority of their lives. Each pig farm will tend to adapt to the local conditions and food supplies and fit their practices to their specific situation.

The following factors can influence the type of pig farms in any given region:

- Available food supply suitable for pigs

- The ability to deal with manure or other outputs from the pig operation

- Local beliefs or traditions, including religion

- The breed or type of pig available to the farm

- Local diseases or conditions that affect pig growth or fecundity

- Local requirements, including government zoning and/or land use laws

- Local and global market conditions and demand

- Traditional farming styles and methods

Use as Food

Almost all of the pig can be used as food. Preparations of pig parts into specialties include: sausage, bacon, gammon, ham, skin into pork scratchings, feet into trotters, head into a meat jelly called head cheese (brawn), and consumption of the liver, chitterlings and blood(blood pudding or black pudding).

Production and Trade

Global pig stocks in 2014 (million)	
People's Republic of China	474.1
United States	67.7
Brazil	37.9
Germany	28.3
Denmark	28.1
Vietnam	26.8
Spain	26.6
Russia	19.1
Mexico	16.1
Myanmar	13.9
World total	**986.6**
Source: *UN Food & Agriculture Organisation (FAO)*	

Pigs are farmed in many countries, though the main consuming countries are in Asia, meaning there is a significant international and even intercontinental trade in live and slaughtered pigs. Despite having the world's largest herd, China is a net importer of pigs, and has been increasing its imports during its economic development. The largest exporters of pigs are the United States,

European Union, and Canada. As an example, more than half of Canadian production (22.8 million pigs) in 2008 was exported, going to 143 countries. Older pigs will consume eleven to nineteen litres (three to five gallons) of water per day.

Relationship Between Handlers and Pigs

The way in which a stockperson interacts with pigs affects animal welfare which in some circumstances can correlate with production measures. Many routine interactions can cause fear, which can result in stress and decreased production.

There are various methods of handling pigs which can be separated into those which lead to positive or negative reactions by the animals. These reactions are based on how the pigs interpret a handler's behavior.

Negative Interactions

Many negative interactions with pigs arise from stockpeople dealing with large numbers of pigs. Because of this, many handlers can become complacent about animal welfare and fail to ensure positive interactions with pigs. Negative interactions include overly-heavy tactile interactions (slaps, punches, kicks and bites), the use of electric goads and fast movements. These can result in fear in the animals, which can develop into stress. Overly-heavy tactile interactions can cause increased basal cortisol levels (a "stress" hormone). Negative interactions that cause fear mean the escape reactions of the pigs can be extremely vigorous, thereby risking injury to both stock and handlers. Stress can result in immunosuppression, leading to an increased susceptibility to disease. Studies have shown that these negative handling techniques result in an overall reduction in growth rates of pigs.

Positive Interactions

Various interactions can be considered either positive or neutral. Neutral interactions are considered positive because, in conjunction with positive interactions, they contribute to an overall non-negative relationship between a stockperson and the stock. Pigs are often fearful of fast movements. When entering a pen, it is good practice for a stockperson to enter with slow and deliberate movements. These minimize fear and therefore reduce stress. Pigs are very curious animals. Allowing the pigs to approach and smell whilst patting or resting a hand on the pig's back are examples of positive behavior. Pigs also respond positively to verbal interaction. Minimising fear of humans allow handlers to perform husbandry practices in a safer and more efficient manner. By reducing stress, stock are more comfortable to feed when near handlers, resulting in increased productivity.

Prohand for pigs is a training program that teaches handlers to interact with pigs in a way that promotes safe handling. It promotes the development of positive behaviors and elimination of negative behaviors. This program has been seen to improve productivity without any capital investment.

Pig Farming Terminology

Pigs are extensively farmed, and therefore the terminology is well developed:

- Pig, hog or swine, the species as a whole, or any member of it. The singular of "swine" is the same as the plural.

- Shoat, piglet or (where the species is called "hog") pig, unweaned young pig, or any immature pig.

- Sucker, a pig between birth and weaning.

- Weaner, a young pig recently separated from the sow.

- Runt, an unusually small and weak piglet, often one in a litter.

- Boar or hog, male pig of breeding age.

- Barrow, male pig castrated before puberty.

- Stag, male pig castrated later in life (an older boar after castration).

- Gilt, young female not yet mated, or not yet farrowed, or after only one litter (depending on local usage).

- Sow, breeding female, or female after first or second litter.

- Suckling pig, a piglet slaughtered for its tender meat.

- Feeder pig, a weaned gilt or barrow weighing between 18 kg (40 lb) and 37 kg (82 lb) at 6 to 8 weeks of age that is sold to be finished for slaughter.

- Porker, market pig between 30 kg (66 lb) and about 54 kg (119 lb) dressed weight.

- Baconer, a market pig between 65 kg (143 lb) and 80 kg (180 lb) dressed weight. The maximum weight can vary between processors.

- Grower, a pig between weaning and sale or transfer to the breeding herd, sold for slaughter or killed for rations.

- Finisher, a grower pig over 70 kg (150 lb) liveweight.

- Butcher hog, a pig of approximately 100 kg (220 lb), ready for the market. In some market (Italy) the final weight of butcher pig is in the 180 kg (400 lb) range. This to have hind legs suitable to produce cured ham.

- Backfatter, cull breeding pig sold for meat; usually refers specifically to a cull sow, but is sometimes used in reference to boars.

Groups

- Herd, a group of pigs, or all the pigs on a farm or in a region.

- Sounder, a small group of pigs (or wild boar) foraging in woodland

Pig Parts

- Trotters, the hooves of pigs (they have four hoofed toes, walking mainly on the larger central two).

Biology

- In pig, pregnant.

- Farrowing, giving birth.

- Hogging, a sow when on heat (during estrus).

Housing

- Sty, a small pig-house, usually with an outdoor run or a pig confinement.

- Pig-shed, a larger pig-house.

- Ark, a low semi circular field-shelter for pigs

- Curtain-barn, a long, open building with curtains on the long sides of the barn. This increases ventilation on hot, humid summer days.

References

- Diamond, J.; Bellwood, P. (2003). "Farmers and Their Languages: The First Expansions". Science. 300 (5619): 597–603. Bibcode:2003Sci...300..597D. doi:10.1126/science.1078208. PMID 12714734.

- Graeme Barker (25 March 2009). The Agricultural Revolution in Prehistory: Why did Foragers become Farmers?. Oxford University Press. ISBN 978-0-19-955995-4. Retrieved 15 August 2012.

- "Anna Creek Station". Wrightsair. Retrieved February 17, 2012. Anna Creek Station is well known as the largest cattle station in the world, covering an area of 24,000 sq. km

- "RSS Text Size Print Share This Home / news / opinion / editorial / Taxpayers Get a Break From Prison Farms". The News & Advance. August 28, 2008. Retrieved February 18, 2012.

- Hoppe, Robert A. and Penni Korb. (2013). Characteristics of Women Farm Operators and Their Farms. Washington, D.C.: U.S. Department of Agriculture, Economic Research Service.

- "Facts on Farmworkers in the United States" (PDF). Cornell University. 2001. Archived from the original (PDF) on December 7, 2006. Retrieved February 17, 2012.

- "Corn Acres Expected to Soar in 2007, USDA Says". Newsroom. Washington: U.S. Department of Agriculture - National Agricultural Statistics Service. March 30, 2007. Retrieved February 18, 2012.

Types of Farming

Farming can be further classified into individual branches of study like organic farming, intensive, extensive farming, etc. which are listed in the following chapter. These distinct types of farming are dealt with great detail in this chapter so that it provides the readers with a comprehensive account of this vast field.

Organic Farming

Vegetables from organic farming.

Organic farming is an alternative agricultural system which originated early in the 20th century in reaction to rapidly changing farming practices. Organic agriculture continues to be developed by various organic agriculture organizations today. It relies on fertilizers of organic origin such as compost, manure, green manure, and bone meal and places emphasis on techniques such as crop rotation and companion planting. Biological pest control, mixed cropping and the fostering of insect predators are encouraged. In general, organic standards are designed to allow the use of naturally occurring substances while prohibiting or strictly limiting synthetic substances.

For instance, naturally occurring pesticides such as pyrethrin and rotenone are permitted, while synthetic fertilizers and pesticides are generally prohibited. Synthetic substances that are allowed include, for example, copper sulfate, elemental sulfur and Ivermectin. Genetically modified organisms, nanomaterials, human sewage sludge, plant growth regulators, hormones, and antibiotic use in livestock husbandry are prohibited. Reasons for advocation of organic farming include real or perceived advantages in sustainability, openness, independence, health, food security, and food safety, although the match between perception and reality is continually challenged.

Organic agricultural methods are internationally regulated and legally enforced by many nations, based in large part on the standards set by the International Federation of Organic Agriculture Movements (IFOAM), an international umbrella organization for organic farming organizations established in 1972. Organic agriculture can be defined as:

an integrated farming system that strives for sustainability, the enhancement of soil fertility and biological diversity whilst, with rare exceptions, prohibiting synthetic pesticides, antibiotics, synthetic fertilizers, genetically modified organisms, and growth hormones.

Since 1990 the market for organic food and other products has grown rapidly, reaching $63 billion worldwide in 2012. This demand has driven a similar increase in organically managed farmland that grew from 2001 to 2011 at a compounding rate of 8.9% per annum. As of 2011, approximately 37,000,000 hectares (91,000,000 acres) worldwide were farmed organically, representing approximately 0.9 percent of total world farmland.

History

Agriculture was practiced for thousands of years without the use of artificial chemicals. Artificial fertilizers were first created during the mid-19th century. These early fertilizers were cheap, powerful, and easy to transport in bulk. Similar advances occurred in chemical pesticides in the 1940s, leading to the decade being referred to as the 'pesticide era'. These new agricultural techniques, while beneficial in the short term, had serious longer term side effects such as soil compaction, erosion, and declines in overall soil fertility, along with health concerns about toxic chemicals entering the food supply. In the late 1800s and early 1900s, soil biology scientists began to seek ways to remedy these side effects while still maintaining higher production.

Biodynamic agriculture was the first modern system of agriculture to focus exclusively on organic methods. Its development began in 1924 with a series of eight lectures on agriculture given by Rudolf Steiner. These lectures, the first known presentation of what later came to be known as organic agriculture, were held in response to a request by farmers who noticed degraded soil conditions and a deterioration in the health and quality of crops and livestock resulting from the use of chemical fertilizers. The one hundred eleven attendees, less than half of whom were farmers, came from six countries, primarily Germany and Poland. The lectures were published in November 1924; the first English translation appeared in 1928 as *The Agriculture Course*.

In 1921, Albert Howard and his wife Gabrielle Howard, accomplished botanists, founded an Institute of Plant Industry to improve traditional farming methods in India. Among other things, they brought improved implements and improved animal husbandry methods from their scientific training; then by incorporating aspects of the local traditional methods, developed protocalls for the rotation of crops, erosion prevention techniques, and the systematic use of composts and manures. Stimulated by these experiences of traditional farming, when Albert Howard returned to Britain in the early 1930s he began to promulgate a system of natural agriculture.

In July 1939, Ehrenfried Pfeiffer, the author of the standard work on biodynamic agriculture (*Bio-Dynamic Farming and Gardening*), came to the UK at the invitation of Walter James, 4th Baron Northbourne as a presenter at the Betteshanger Summer School and Conference on Biodynamic Farming at Northbourne's farm in Kent. One of the chief purposes of the conference was to bring together the proponents of various approaches to organic agriculture in order that they might cooperate within a larger movement. Howard attended the conference, where he met Pfeiffer. In the following year, Northbourne published his manifesto of organic farming, *Look to the Land*, in which he coined the term "organic farming." The Betteshanger conference has been described as the 'missing link' between biodynamic agriculture and other forms of organic farming.

In 1940 Howard published his *An Agricultural Testament*. In this book he adopted Northbourne's terminology of "organic farming." Howard's work spread widely, and he became known as the "father of organic farming" for his work in applying scientific knowledge and principles to various traditional and natural methods. In the United States J.I. Rodale, who was keenly interested both in Howard's ideas and in biodynamics, founded in the 1940s both a working organic farm for trials and experimentation, The Rodale Institute, and the Rodale Press to teach and advocate organic methods to the wider public. These became important influences on the spread of organic agriculture. Further work was done by Lady Eve Balfour in the United Kingdom, and many others across the world.

Increasing environmental awareness in the general population in modern times has transformed the originally supply-driven organic movement to a demand-driven one. Premium prices and some government subsidies attracted farmers. In the developing world, many producers farm according to traditional methods that are comparable to organic farming, but not certified, and that may not include the latest scientific advancements in organic agriculture. In other cases, farmers in the developing world have converted to modern organic methods for economic reasons.

Terminology

Biodynamic agriculturists, who based their work on Steiner's spiritually-oriented anthroposophy, used the term "organic" to indicate that a farm should be viewed as a living organism, in the sense of the following quotation:

> "An organic farm, properly speaking, is not one that uses certain methods and substances and avoids others; it is a farm whose structure is formed in imitation of the structure of a natural system that has the integrity, the independence and the benign dependence of an organism"

> — *Wendell Berry, "The Gift of Good Land"*

The use of "organic" popularized by Howard and Rodale, on the other hand, refers more narrowly to the use of organic matter derived from plant compost and animal manures to improve the humus content of soils, grounded in the work of early soil scientists who developed what was then called "humus farming." Since the early 1940s the two camps have tended to merge.

Methods

Organic cultivation of mixed vegetables in Capay, California. Note the hedgerow in the background.

"Organic agriculture is a production system that sustains the health of soils, ecosystems and people. It relies on ecological processes, biodiversity and cycles adapted to local conditions, rather than the use of inputs with adverse effects. Organic agriculture combines tradition, innovation and science to benefit the shared environment and promote fair relationships and a good quality of life for all involved..."

— *International Federation of Organic Agriculture Movements*

Organic farming methods combine scientific knowledge of ecology and modern technology with traditional farming practices based on naturally occurring biological processes. Organic farming methods are studied in the field of agroecology. While conventional agriculture uses synthetic pesticides and water-soluble synthetically purified fertilizers, organic farmers are restricted by regulations to using natural pesticides and fertilizers. An example of a natural pesticide is pyrethrin, which is found naturally in the Chrysanthemum flower. The principal methods of organic farming include crop rotation, green manures and compost, biological pest control, and mechanical cultivation. These measures use the natural environment to enhance agricultural productivity: legumes are planted to fix nitrogen into the soil, natural insect predators are encouraged, crops are rotated to confuse pests and renew soil, and natural materials such as potassium bicarbonate and mulches are used to control disease and weeds. Genetically modified seeds and animals are excluded.

While organic is fundamentally different from conventional because of the use of carbon based fertilizers compared with highly soluble synthetic based fertilizers and biological pest control instead of synthetic pesticides, organic farming and large-scale conventional farming are not entirely mutually exclusive. Many of the methods developed for organic agriculture have been borrowed by more conventional agriculture. For example, Integrated Pest Management is a multifaceted strategy that uses various organic methods of pest control whenever possible, but in conventional farming could include synthetic pesticides only as a last resort.

Crop Diversity

Organic farming encourages Crop diversity. The science of agroecology has revealed the benefits of polyculture (multiple crops in the same space), which is often employed in organic farming. Planting a variety of vegetable crops supports a wider range of beneficial insects, soil microorganisms, and other factors that add up to overall farm health. Crop diversity helps environments thrive and protects species from going extinct.

Soil Management

Organic farming relies heavily on the natural breakdown of organic matter, using techniques like green manure and composting, to replace nutrients taken from the soil by previous crops. This biological process, driven by microorganisms such as mycorrhiza, allows the natural production of nutrients in the soil throughout the growing season, and has been referred to as *feeding the soil to feed the plant*. Organic farming uses a variety of methods to improve soil fertility, including crop rotation, cover cropping, reduced tillage, and application of compost. By reducing tillage, soil is not inverted and exposed to air; less carbon is lost to the atmosphere resulting in more soil organic carbon. This has an added benefit of carbon sequestration, which can reduce green house gases and help reverse climate change.

Plants need nitrogen, phosphorus, and potassium, as well as micronutrients and symbiotic relationships with fungi and other organisms to flourish, but getting enough nitrogen, and particularly synchronization so that plants get enough nitrogen at the right time (when plants need it most), is a challenge for organic farmers. Crop rotation and green manure ("cover crops") help to provide nitrogen through legumes (more precisely, the *Fabaceae* family), which fix nitrogen from the atmosphere through symbiosis with rhizobial bacteria. Intercropping, which is sometimes used for insect and disease control, can also increase soil nutrients, but the competition between the legume and the crop can be problematic and wider spacing between crop rows is required. Crop residues can be ploughed back into the soil, and different plants leave different amounts of nitrogen, potentially aiding synchronization. Organic farmers also use animal manure, certain processed fertilizers such as seed meal and various mineral powders such as rock phosphate and green sand, a naturally occurring form of potash that provides potassium. Together these methods help to control erosion. In some cases pH may need to be amended. Natural pH amendments include lime and sulfur, but in the U.S. some compounds such as iron sulfate, aluminum sulfate, magnesium sulfate, and soluble boron products are allowed in organic farming.

Mixed farms with both livestock and crops can operate as ley farms, whereby the land gathers fertility through growing nitrogen-fixing forage grasses such as white clover or alfalfa and grows cash crops or cereals when fertility is established. Farms without livestock ("stockless") may find it more difficult to maintain soil fertility, and may rely more on external inputs such as imported manure as well as grain legumes and green manures, although grain legumes may fix limited nitrogen because they are harvested. Horticultural farms that grow fruits and vegetables in protected conditions often relay even more on external inputs.

Biological research into soil and soil organisms has proven beneficial to organic farming. Varieties of bacteria and fungi break down chemicals, plant matter and animal waste into productive soil nutrients. In turn, they produce benefits of healthier yields and more productive soil for future crops. Fields with less or no manure display significantly lower yields, due to decreased soil microbe community. Increased manure improves biological activity, providing a healthier, more arable soil system and higher yields.

Weed Management

Organic weed management promotes weed suppression, rather than weed elimination, by enhancing crop competition and phytotoxic effects on weeds. Organic farmers integrate cultural, biological, mechanical, physical and chemical tactics to manage weeds without synthetic herbicides.

Organic standards require rotation of annual crops, meaning that a single crop cannot be grown in the same location without a different, intervening crop. Organic crop rotations frequently include weed-suppressive cover crops and crops with dissimilar life cycles to discourage weeds associated with a particular crop. Research is ongoing to develop organic methods to promote the growth of natural microorganisms that suppress the growth or germination of common weeds.

Other cultural practices used to enhance crop competitiveness and reduce weed pressure include selection of competitive crop varieties, high-density planting, tight row spacing, and late planting into warm soil to encourage rapid crop germination.

Mechanical and physical weed control practices used on organic farms can be broadly grouped as:

- Tillage - Turning the soil between crops to incorporate crop residues and soil amendments; remove existing weed growth and prepare a seedbed for planting; turning soil after seeding to kill weeds, including cultivation of row crops;

- Mowing and cutting - Removing top growth of weeds;

- Flame weeding and thermal weeding - Using heat to kill weeds; and

- Mulching - Blocking weed emergence with organic materials, plastic films, or landscape fabric.

Some critics, citing work published in 1997 by David Pimentel of Cornell University, which described an epidemic of soil erosion worldwide, have raised concerned that tillage contribute to the erosion epidemic. The FAO and other organizations have advocated a 'no-till' approach to both conventional and organic farming, and point out in particular that crop rotation techniques used in organic farming are excellent no-till approaches. A study published in 2005 by Pimentel and colleagues confirmed that 'Crop rotations and cover cropping (green manure) typical of organic agriculture reduce soil erosion, pest problems, and pesticide use.' Some naturally sourced chemicals are allowed for herbicidal use. These include certain formulations of acetic acid (concentrated vinegar), corn gluten meal, and essential oils. A few selective bioherbicides based on fungal pathogens have also been developed. At this time, however, organic herbicides and bioherbicides play a minor role in the organic weed control toolbox.

Weeds can be controlled by grazing. For example, geese have been used successfully to weed a range of organic crops including cotton, strawberries, tobacco, and corn, reviving the practice of keeping cotton patch geese, common in the southern U.S. before the 1950s. Similarly, some rice farmers introduce ducks and fish to wet paddy fields to eat both weeds and insects.

Controlling Other Organisms

Chloroxylon is used for Pest Management in Organic Rice Cultivation in Chhattisgarh, India

Organisms aside from weeds that cause problems on organic farms include arthropods (e.g., insects, mites), nematodes, fungi and bacteria. Organic practices include, but are not limited to:

- encouraging predatory beneficial insects to control pests by serving them nursery plants and/or an alternative habitat, usually in a form of a shelterbelt, hedgerow, or beetle bank;

- encouraging beneficial microorganisms;

- rotating crops to different locations from year to year to interrupt pest reproduction cycles;

- planting companion crops and pest-repelling plants that discourage or divert pests;

- using row covers to protect crops during pest migration periods;

- using biologic pesticides and herbicides

- using stale seed beds to germinate and destroy weeds before planting

- using sanitation to remove pest habitat;

- Using insect traps to monitor and control insect populations.

- Using physical barriers, such as row covers

Examples of predatory beneficial insects include minute pirate bugs, big-eyed bugs, and to a lesser extent ladybugs (which tend to fly away), all of which eat a wide range of pests. Lacewings are also effective, but tend to fly away. Praying mantis tend to move more slowly and eat less heavily. Parasitoid wasps tend to be effective for their selected prey, but like all small insects can be less effective outdoors because the wind controls their movement. Predatory mites are effective for controlling other mites.

Naturally derived insecticides allowed for use on organic farms use include *Bacillus thuringiensis* (a bacterial toxin), pyrethrum (a chrysanthemum extract), spinosad (a bacterial metabolite), neem (a tree extract) and rotenone (a legume root extract). Fewer than 10% of organic farmers use these pesticides regularly; one survey found that only 5.3% of vegetable growers in California use rotenone while 1.7% use pyrethrum. These pesticides are not always more safe or environmentally friendly than synthetic pesticides and can cause harm. The main criterion for organic pesticides is that they are naturally derived, and some naturally derived substances have been controversial. Controversial natural pesticides include rotenone, copper, nicotine sulfate, and pyrethrums Rotenone and pyrethrum are particularly controversial because they work by attacking the nervous system, like most conventional insecticides. Rotenone is extremely toxic to fish and can induce symptoms resembling Parkinson's disease in mammals. Although pyrethrum (natural pyrethrins) is more effective against insects when used with piperonyl butoxide (which retards degradation of the pyrethrins), organic standards generally do not permit use of the latter substance.

Naturally derived fungicides allowed for use on organic farms include the bacteria *Bacillus subtilis* and *Bacillus pumilus*; and the fungus *Trichoderma harzianum*. These are mainly effective for diseases affecting roots. Compost tea contains a mix of beneficial microbes, which may attack or out-compete certain plant pathogens, but variability among formulations and preparation methods may contribute to inconsistent results or even dangerous growth of toxic microbes in compost teas.

Some naturally derived pesticides are not allowed for use on organic farms. These include nicotine sulfate, arsenic, and strychnine.

Synthetic pesticides allowed for use on organic farms include insecticidal soaps and horticultural oils for insect management; and Bordeaux mixture, copper hydroxide and sodium bicarbonate for managing fungi. Copper sulfate and Bordeaux mixture (copper sulfate plus lime), approved for organic use in various jurisdictions, can be more environmentally problematic than some synthetic fungicides dissallowed in organic farming Similar concerns apply to copper hydroxide. Repeated application of copper sulfate or copper hydroxide as a fungicide may eventually result in copper accumulation to toxic levels in soil, and admonitions to avoid excessive accumulations of copper in soil appear in various organic standards and elsewhere. Environmental concerns for several kinds of biota arise at average rates of use of such substances for some crops. In the European Union, where replacement of copper-based fungicides in organic agriculture is a policy priority, research is seeking alternatives for organic production.

Livestock

For livestock like these healthy cows vaccines play an important part in animal health since antibiotic therapy is prohibited in organic farming

Raising livestock and poultry, for meat, dairy and eggs, is another traditional farming activity that complements growing. Organic farms attempt to provide animals with natural living conditions and feed. Organic certification verifies that livestock are raised according to the USDA organic regulations throughout their lives. These regulations include the requirement that all animal feed must be certified organic.

Organic livestock may be, and must be, treated with medicine when they are sick, but drugs cannot be used to promote growth, their feed must be organic, and they must be pastured.

Also, horses and cattle were once a basic farm feature that provided labor, for hauling and plowing, fertility, through recycling of manure, and fuel, in the form of food for farmers and other animals. While today, small growing operations often do not include livestock, domesticated animals are a desirable part of the organic farming equation, especially for true sustainability, the ability of a farm to function as a self-renewing unit.

Genetic Modification

A key characteristic of organic farming is the rejection of genetically engineered plants and animals. On 19 October 1998, participants at IFOAM's 12th Scientific Conference issued the Mar del Plata Declaration, where more than 600 delegates from over 60 countries voted unanimously to exclude the use of genetically modified organisms in food production and agriculture.

Although opposition to the use of any transgenic technologies in organic farming is strong, agricultural researchers Luis Herrera-Estrella and Ariel Alvarez-Morales continue to advocate integration of transgenic technologies into organic farming as the optimal means to sustainable agriculture, particularly in the developing world, as does author and scientist Pamela Ronald, who views this kind of biotechnology as being consistent with organic principles.

Although GMOs are excluded from organic farming, there is concern that the pollen from genetically modified crops is increasingly penetrating organic and heirloom seed stocks, making it difficult, if not impossible, to keep these genomes from entering the organic food supply. Differing regulations among countries limits the availability of GMOs to certain countries, as described in the article on regulation of the release of genetic modified organisms.

Tools

Organic farmers use a number of traditional farm tools to do farming. Due to the goals of sustainability in organic farming, organic farmers try to minimize their reliance on fossil fuels. In the developing world on small organic farms tools are normally constrained to hand tools and diesel powered water pumps. Some organic farmers make use of renewable energy on the farm and can even make use of agrivoltaics or other onsite colocation of power production and agriculture. A recent study evaluated the use of open-source 3-D printers (called RepRaps using a bioplastic polylactic acid (PLA) on organic farms. PLA is a strong biodegradable and recyclable thermoplastic appropriate for a range of representative products in five categories of prints: handtools, food processing, animal management, water management and hydroponics. Such open source hardware is attractive to all types of small farmers as it provides control for farmers over their own equipment; this is exemplified by Open Source Ecology, Farm Hack and Farmbot.io.

Standards

Standards regulate production methods and in some cases final output for organic agriculture. Standards may be voluntary or legislated. As early as the 1970s private associations certified organic producers. In the 1980s, governments began to produce organic production guidelines. In the 1990s, a trend toward legislated standards began, most notably with the 1991 EU-Eco-regulation developed for European Union, which set standards for 12 countries, and a 1993 UK program. The EU's program was followed by a Japanese program in 2001, and in 2002 the U.S. created the National Organic Program (NOP). As of 2007 over 60 countries regulate organic farming (IFOAM 2007:11). In 2005 IFOAM created the Principles of Organic Agriculture, an international guideline for certification criteria. Typically the agencies accredit certification groups rather than individual farms.

Organic production materials used in and foods are tested independently by the Organic Materials Review Institute.

Composting

Using manure as a fertiliser risks contaminating food with animal gut bacteria, including pathogenic strains of E. coli that have caused fatal poisoning from eating organic food. To combat this risk, USDA organic standards require that manure must be sterilized through high temperature thermophilic composting. If raw animal manure is used, 120 days must pass before the crop is harvested if the final product comes into direct contact with the soil. For products that don't directly contact soil, 90 days must pass prior to harvest.

Economics

The economics of organic farming, a subfield of agricultural economics, encompasses the entire process and effects of organic farming in terms of human society, including social costs, opportunity costs, unintended consequences, information asymmetries, and economies of scale. Although the scope of economics is broad, agricultural economics tends to focus on maximizing yields and efficiency at the farm level. Economics takes an anthropocentric approach to the value of the natural world: biodiversity, for example, is considered beneficial only to the extent that it is valued by people and increases profits. Some entities such as the European Union subsidize organic farming, in large part because these countries want to account for the externalities of reduced water use, reduced water contamination, reduced soil erosion, reduced carbon emissions, increased biodiversity, and assorted other benefits that result from organic farming.

Traditional organic farming is labor and knowledge-intensive whereas conventional farming is capital-intensive, requiring more energy and manufactured inputs.

Organic farmers in California have cited marketing as their greatest obstacle.

Geographic Producer Distribution

The markets for organic products are strongest in North America and Europe, which as of 2001 are estimated to have $6 and $8 billion respectively of the $20 billion global market. As of 2007 Australasia has 39% of the total organic farmland, including Australia's 1,180,000 hectares (2,900,000 acres) but 97 percent of this land is sprawling rangeland (2007:35). US sales are 20x as much. Europe farms 23 percent of global organic farmland (6,900,000 ha (17,000,000 acres)), followed by Latin America with 19 percent (5.8 million hectares - 14.3 million acres). Asia has 9.5 percent while North America has 7.2 percent. Africa has 3 percent.

Besides Australia, the countries with the most organic farmland are Argentina (3.1 million hectares - 7.7 million acres), China (2.3 million hectares - 5.7 million acres), and the United States (1.6 million hectares - 4 million acres). Much of Argentina's organic farmland is pasture, like that of Australia (2007:42). Spain, Germany, Brazil (the world's largest agricultural exporter), Uruguay, and the UK follow the United States in the amount of organic land (2007:26).

In the European Union (EU25) 3.9% of the total utilized agricultural area was used for organic production in 2005. The countries with the highest proportion of organic land were Austria (11%) and Italy (8.4%), followed by the Czech Republic and Greece (both 7.2%). The lowest figures were shown for Malta (0.1%), Poland (0.6%) and Ireland (0.8%). In 2009, the proportion of organic land in the EU grew to 4.7%. The countries with highest share of agricultural land were Liechtenstein (26.9%), Austria (18.5%) and Sweden (12.6%). 16% of all farmers in Austria produced organically in 2010. By the same year the proportion of organic land increased to 20%.: In 2005 168,000 ha (415,000 ac) of land in Poland was under organic management. In 2012, 288,261 hectares (712,308 acres) were under organic production, and there were about 15,500 organic farmers; retail sales of organic products were EUR 80 million in 2011. As of 2012 organic exports were part of the government's economic development strategy.

After the collapse of the Soviet Union in 1991, agricultural inputs that had previously been purchased from Eastern bloc countries were no longer available in Cuba, and many Cuban farms converted to organic methods out of necessity. Consequently, organic agriculture is a mainstream practice in Cuba, while it remains an alternative practice in most other countries. Cuba's organic strategy includes development of genetically modified crops; specifically corn that is resistant to the palomilla moth

Growth

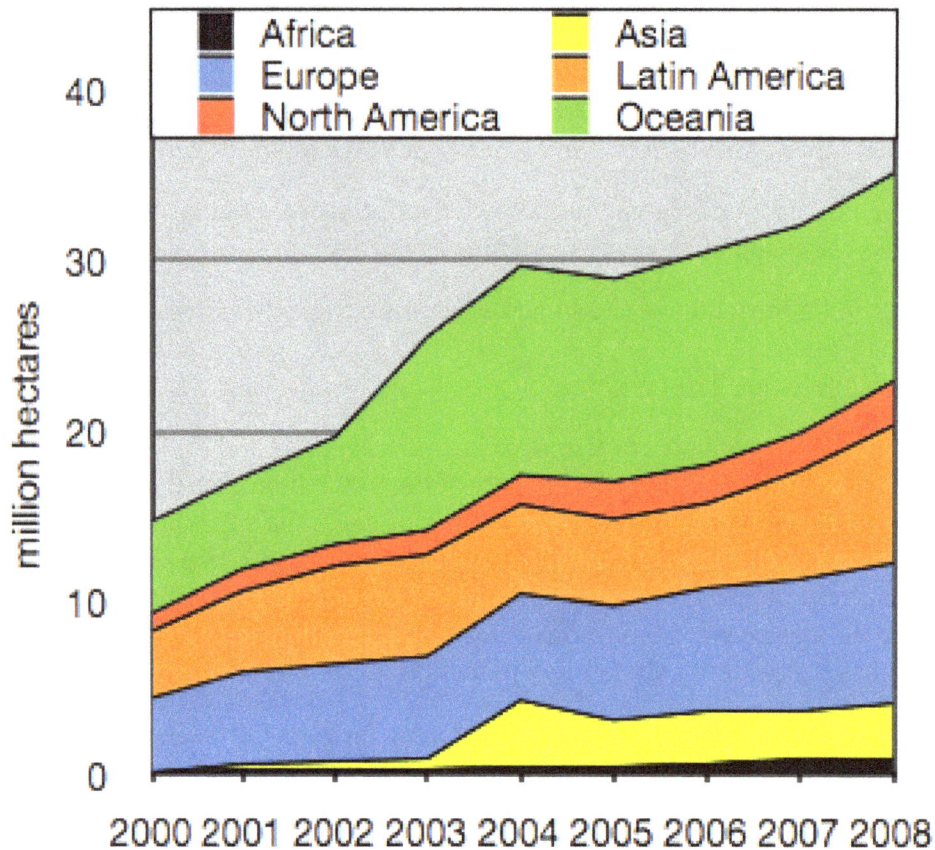

Organic farmland by world region (2000-2008)

In 2001, the global market value of certified organic products was estimated at USD $20 billion. By 2002, this was USD $23 billion and by 2007 more than USD $46 billion. By 2014, retail sales of organic products reached USD $80 billion worldwide. North America and Europe accounted for more than 90% of all organic product sales.

Organic agricultural land increased almost fourfold in 15 years, from 11 million hectares in 1999 to 43.7 million hectares in 2014. Between 2013 and 2014, organic agricultural land grew by 500,000 hectares worldwide, increasing in every region except Latin America. During this time period, Europe's organic farmland increased 260,000 hectares to 11.6 million total (+2.3%), Asia's increased 159,000 hectares to 3.6 million total (+4.7%), Africa's increased 54,000 hectares to 1.3 million total (+4.5%), and North America's increased 35,000 hectares to 3.1 million total (+1.1%). As of 2014, the country with the most organic land was Australia (17.2 million hectares), followed by Argentina (3.1 million hectares), and the United States (2.2 million hectares).

In 2013, the number of organic producers grew by almost 270,000, or more than 13%. By 2014, there were a reported 2.3 million organic producers in the world. Most of the total global increase took place in the Philippines, Peru, China, and Thailand. Overall, the majority of all organic producers are in India (650,000 in 2013), Uganda (190,552 in 2014), Mexico (169,703 in 2013) and the Philippines (165,974 in 2014).

Productivity

Studies comparing yields have had mixed results. These differences among findings can often be attributed to variations between study designs including differences in the crops studied and the methodology by which results were gathered.

A 2012 meta-analysis found that productivity is typically lower for organic farming than conventional farming, but that the size of the difference depends on context and in some cases may be very small. While organic yields can be lower than conventional yields, another meta-analysis published in Sustainable Agriculture Research in 2015, concluded that certain organic on-farm practices could help narrow this gap. Timely weed management and the application of manure in conjunction with legume forages/cover crops were shown to have positive results in increasing organic corn and soybean productivity. More experienced organic farmers were also found to have higher yields than other organic farmers who were just starting out.

Another meta-analysis published in the journal Agricultural Systems in 2011 analyzed 362 datasets and found that organic yields were on average 80% of conventional yields. The author's found that there are relative differences in this yield gap based on crop type with crops like soybeans and rice scoring higher than the 80% average and crops like wheat and potato scoring lower. Across global regions, Asia and Central Europe were found to have relatively higher yields and Northern Europe relatively lower than the average.

A 2007 study compiling research from 293 different comparisons into a single study to assess the overall efficiency of the two agricultural systems has concluded that "organic methods could produce enough food on a global per capita basis to sustain the current human population, and potentially an even larger population, without increasing the agricultural land base." The researchers also found that while in developed countries, organic systems on average produce 92% of the yield produced by conventional agriculture, organic systems produce 80% more than conventional farms in developing countries, because the materials needed for organic farming are more accessible than synthetic farming materials to farmers in some poor countries. This study was strongly contested by another study published in 2008, which stated, and was entitled, "Organic agriculture cannot feed the world" and said that the 2007 came up with "a major overestimation of the productivity of OA" "because data are misinterpreted and calculations accordingly are erroneous." Additional research needs to be conducted in the future to further clarify these claims.

Long Term Studies

A study published in 2005 compared conventional cropping, organic animal-based cropping, and organic legume-based cropping on a test farm at the Rodale Institute over 22 years. The study found that "the crop yields for corn and soybeans were similar in the organic animal, organic legume, and conventional farming systems". It also found that "significantly less fossil energy was expended to produce corn in the Rodale Institute's organic animal and organic legume systems than in the conventional production system. There was little difference in energy input between the different treatments for producing soybeans. In the organic systems, synthetic fertilizers and pesticides were generally not used". As of 2013 the Rodale study was ongoing and a thirty-year anniversary report was published by Rodale in 2012.

A long-term field study comparing organic/conventional agriculture carried out over 21 years in Switzerland concluded that "Crop yields of the organic systems averaged over 21 experimental years at 80% of the conventional ones. The fertilizer input, however, was 34 – 51% lower, indicating an efficient production. The organic farming systems used 20 – 56% less energy to produce a crop unit and per land area this difference was 36 – 53%. In spite of the considerably lower pesticide input the quality of organic products was hardly discernible from conventional analytically and even came off better in food preference trials and picture creating methods"

Profitability

In the United States, organic farming has been shown to be 2.9 to 3.8 times more profitable for the farmer than conventional farming when prevailing price premiums are taken into account. Globally, organic farming is between 22 and 35 percent more profitable for farmers than conventional methods, according to a 2015 meta-analysis of studies conducted across five continents.

The profitability of organic agriculture can be attributed to a number of factors. First, organic farmers do not rely on synthetic fertilizer and pesticide inputs, which can be costly. In addition, organic foods currently enjoy a price premium over conventionally produced foods, meaning that organic farmers can often get more for their yield.

The price premium for organic food is an important factor in the economic viability of organic farming. In 2013 there was a 100% price premium on organic vegetables and a 57% price premium for organic fruits. These percentages are based on wholesale fruit and vegetable prices, available through the United States Department of Agriculture's Economic Research Service. Price premiums exist not only for organic versus nonorganic crops, but may also vary depending on the venue where the product is sold: farmers markets, grocery stores, or wholesale to restaurants. For many producers, direct sales at farmers markets are most profitable because the farmer receives the entire markup, however this is also the most time and labor-intensive approach.

There have been signs of organic price premiums narrowing in recent years, which lowers the economic incentive for farmers to convert to or maintain organic production methods. Data from 22 years of experiments at the Rodale Institute found that, based on the current yields and production costs associated with organic farming in the United States, a price premium of only 10% is required to achieve parity with conventional farming. A separate study found that on a global scale, price premiums of only 5-7% percent were needed to break even with conventional methods. Without the price premium, profitability for farmers is mixed.

For markets and supermarkets organic food is profitable as well, and is generally sold at significantly higher prices than non-organic food.

Energy Efficiency

In the most recent assessments of the energy efficiency of organic versus conventional agriculture, results have been mixed regarding which form is more carbon efficient. Organic farm systems have more often than not been found to be more energy efficient, however, this is not always the case. More than anything, results tend to depend upon crop type and farm size.

A comprehensive comparison of energy efficiency in grain production, produce yield, and animal husbandry concluded that organic farming had a higher yield per unit of energy over the vast majority of the crops and livestock systems. For example, two studies - both comparing organically-versus conventionally-farmed apples - declare contradicting results, one saying organic farming is more energy efficient, the other saying conventionally is more efficient.

It has generally been found that the labor input per unit of yield was higher for organic systems compared with conventional production.

Sales and Marketing

Most sales are concentrated in developed nations. In 2008, 69% of Americans claimed to occasionally buy organic products, down from 73% in 2005. One theory for this change was that consumers were substituting "local" produce for "organic" produce.

Distributors

The USDA requires that distributors, manufacturers, and processors of organic products be certified by an accredited state or private agency. In 2007, there were 3,225 certified organic handlers, up from 2,790 in 2004.

Organic handlers are often small firms; 48% reported sales below $1 million annually, and 22% between $1 and $5 million per year. Smaller handlers are more likely to sell to independent natural grocery stores and natural product chains whereas large distributors more often market to natural product chains and conventional supermarkets, with a small group marketing to independent natural product stores. Some handlers work with conventional farmers to convert their land to organic with the knowledge that the farmer will have a secure sales outlet. This lowers the risk for the handler as well as the farmer. In 2004, 31% of handlers provided technical support on organic standards or production to their suppliers and 34% encouraged their suppliers to transition to organic. Smaller farms often join together in cooperatives to market their goods more effectively.

93% of organic sales are through conventional and natural food supermarkets and chains, while the remaining 7% of U.S. organic food sales occur through farmers' markets, foodservices, and other marketing channels.

Direct-to-Consumer Sales

In the 2012 Census, direct-to-consumer sales equaled $1.3 billion, up from $812 million in 2002, an increase of 60 percent. The number of farms that utilize direct-to-consumer sales was 144,530 in 2012 in comparison to 116,733 in 2002. Direct-to-consumer sales include farmers markets, community supported agriculture (CSA), on-farm stores, and roadside farm stands. Some organic farms also sell products direct to retailer, direct to restaurant and direct to institution. According to the 2008 Organic Production Survey, approximately 7% of organic farm sales went direct-to-consumers, 10% went direct to retailers, and approximately 83% went into wholesale markets. In comparison, only 0.4% of the value of convention agricultural commodities went direct-to-consumers.

While not all products sold at farmer's markets are certified organic, this direct-to-consumer avenue has become increasingly popular in local food distribution and has grown substantially since 1994. In 2014, there were 8,284 farmer's markets in comparison to 3,706 in 2004 and 1,755 in 1994, most of which are found in populated areas such as the Northeast, Midwest, and West Coast.

Labor and Employment

Organic production is more labor-intensive than conventional production. On the one hand, this increased labor cost is one factor that makes organic food more expensive. On the other hand, the increased need for labor may be seen as an "employment dividend" of organic farming, providing more jobs per unit area than conventional systems. The 2011 UNEP Green Economy Report suggests that "[a]n increase in investment in green agriculture is projected to lead to growth in employment of about 60 per cent compared with current levels" and that "green agriculture investments could create 47 million additional jobs compared with BAU2 over the next 40 years." The UNEP also argues that "[b]y greening agriculture and food distribution, more calories per person per day, more jobs and business opportunities especially in rural areas, and market-access opportunities, especially for developing countries, will be available."

World's Food Security

In 2007 the United Nations Food and Agriculture Organization (FAO) said that organic agriculture often leads to higher prices and hence a better income for farmers, so it should be promoted. However, FAO stressed that by organic farming one could not feed the current mankind, even less the bigger future population. Both data and models showed then that organic farming was far from sufficient. Therefore, chemical fertilizers were needed to avoid hunger. Other analysis by many agribusiness executives, agricultural and ecological scientists, and international agriculture experts revealed the opinion that organic farming would not only increase the world's food supply, but might be the only way to eradicate hunger.

FAO stressed that fertilizers and other chemical inputs can much increase the production, particularly in Africa where fertilizers are currently used 90% less than in Asia. For example, in Malawi the yield has been boosted using seeds and fertilizers. FAO also calls for using biotechnology, as it can help smallholder farmers to improve their income and food security.

Also NEPAD, development organization of African governments, announced that feeding Africans and preventing malnutrition requires fertilizers and enhanced seeds.

According to a more recent study in ScienceDigest, organic best management practices shows an average yield only 13% less than conventional. In the world's poorer nations where most of the world's hungry live, and where conventional agriculture's expensive inputs are not affordable by the majority of farmers, adopting organic management actually increases yields 93% on average, and could be an important part of increased food security.

Capacity Building in Developing Countries

Organic agriculture can contribute to ecologically sustainable, socio-economic development, especially in poorer countries. The application of organic principles enables employment of local resources (e.g., local seed varieties, manure, etc.) and therefore cost-effectiveness. Local and international markets for organic products show tremendous growth prospects and offer creative producers and exporters excellent opportunities to improve their income and living conditions.

Organic agriculture is knowledge intensive. Globally, capacity building efforts are underway, including localized training material, to limited effect. As of 2007, the International Federation of Organic Agriculture Movements hosted more than 170 free manuals and 75 training opportunities online.

In 2008 the United Nations Environmental Programme (UNEP) and the United Nations Conference on Trade and Development (UNCTAD) stated that "organic agriculture can be more conducive to food security in Africa than most conventional production systems, and that it is more likely to be sustainable in the long-term" and that "yields had more than doubled where organic, or near-organic practices had been used" and that soil fertility and drought resistance improved.

Millennium Development Goals

The value of organic agriculture (OA) in the achievement of the Millennium Development Goals (MDG), particularly in poverty reduction efforts in the face of climate change, is shown by its contribution to both income and non-income aspects of the MDGs. These benefits are expected to continue in the post-MDG era. A series of case studies conducted in selected areas in Asian countries by the Asian Development Bank Institute (ADBI) and published as a book compilation by ADB in Manila document these contributions to both income and non-income aspects of the MDGs. These include poverty alleviation by way of higher incomes, improved farmers' health owing to less chemical exposure, integration of sustainable principles into rural development policies, improvement of access to safe water and sanitation, and expansion of global partnership for development as small farmers are integrated in value chains.

A related ADBI study also sheds on the costs of OA programs and set them in the context of the costs of attaining the MDGs. The results show considerable variation across the case studies, suggesting that there is no clear structure to the costs of adopting OA. Costs depend on the efficiency of the OA adoption programs. The lowest cost programs were more than ten times less expensive than the highest cost ones. However, further analysis of the gains resulting from OA adoption reveals that the costs per person taken out of poverty was much lower than the estimates of the World Bank, based on income growth in general or based on the detailed costs of meeting some of the more quantifiable MDGs (e.g., education, health, and environment).

Externalities

Agriculture imposes negative externalities (uncompensated costs) upon society through land and other resource use, biodiversity loss, erosion, pesticides, nutrient runoff, water usage, subsidy payments and assorted other problems. Positive externalities include self-reliance, entrepreneurship, respect for nature, and air quality. Organic methods reduce some of these costs. In 2000 uncompensated costs for 1996 reached 2,343 million British pounds or £208 per ha (£84.20/ac). A study of practices in the USA published in 2005 concluded that cropland costs the economy approximately 5 to 16 billion dollars ($30–96/ha - $12–39/ac), while livestock production costs 714 million dollars. Both studies recommended reducing externalities. The 2000 review included reported pesticide poisonings but did not include speculative chronic health effects of pesticides, and the 2004 review relied on a 1992 estimate of the total impact of pesticides.

It has been proposed that organic agriculture can reduce the level of some negative externalities from (conventional) agriculture. Whether the benefits are private or public depends upon the division of property rights.

Several surveys and studies have attempted to examine and compare conventional and organic systems of farming and have found that organic techniques, while not without harm, are less damaging than conventional ones because they reduce levels of biodiversity less than conventional systems do and use less energy and produce less waste when calculated per unit area.

A 2003 to 2005 investigation by the Cranfield University for the Department for Environment Food and Rural Affairs in the UK found that it is difficult to compare the Global Warming Potential (GWP), acidification and eutrophication emissions but "Organic production often results in increased burdens, from factors such as N leaching and N2O emissions", even though primary energy use was less for most organic products. N_2O is always the largest GWP contributor except in tomatoes. However, "organic tomatoes always incur more burdens (except pesticide use)". Some emissions were lower "per area", but organic farming always required 65 to 200% more field area than non-organic farming. The numbers were highest for bread wheat (200+ % more) and potatoes (160% more).

The situation was shown dramatically in a comparison of a modern dairy farm in Wisconsin with one in New Zealand in which the animals grazed extensively. Using total farm emissions per kg milk produced as a parameter, the researchers showed that production of methane from belching was higher in the New Zealand farm, while carbon dioxide production was higher in the Wisconsin farm. Output of nitrous oxide, a gas with an estimated global warming potential 310 times that of carbon dioxide was also higher in the New Zealand farm. Methane from manure handling was similar in the two types of farm. The explanation for the finding relates to the different diets used on these farms, being based more completely on forage (and hence more fibrous) in New Zealand and containing less concentrate than in Wisconsin. Fibrous diets promote a higher proportion of acetate in the gut of ruminant animals, resulting in a higher production of methane that must be released by belching. When cattle are given a diet containing some concentrates (such as corn and soybean meal) in addition to grass and silage, the pattern of ruminal fermentation alters from acetate to mainly propionate. As a result, methane production is reduced. Capper et al. compared the environmental impact of US dairy production in 1944 and 2007. They calculated that the carbon "footprint" per billion kg (2.2 billion lb) of milk produced in 2007 was 37 percent that of equivalent milk production in 1944.

Environmental Impact and Emissions

Researchers at Oxford university analyzed 71 peer-reviewed studies and observed that organic products are sometimes worse for the environment. Organic milk, cereals, and pork generated higher greenhouse gas emissions per product than conventional ones but organic beef and olives had lower emissions in most studies. Usually organic products required less energy, but more land. Per unit of product, organic produce generates higher nitrogen leaching, nitrous oxide emissions, ammonia emissions, eutrophication and acidification potential than when conventionally grown. Other differences were not significant. The researchers concluded "Most of the studies that compared biodiversity in organic and conventional farming demonstrated lower environmental impacts from organic farming." The researchers believe that the ideal outcome would be to develop new systems that consider both the environment, including setting land aside for wildlife and sustainable forestry, and the development of ways to produce the highest yields possible using both conventional and organic methods.

Proponents of organic farming have claimed that organic agriculture emphasizes closed nutrient cycles, biodiversity, and effective soil management providing the capacity to mitigate and even reverse the effects of climate change and that organic agriculture can decrease fossil fuel emissions. "The carbon sequestration efficiency of organic systems in temperate climates is almost double (575-700 kg carbon per ha per year - 510-625 lb/ac/an) that of conventional treatment of soils, mainly owing to the use of grass clovers for feed and of cover crops in organic rotations."

Critics of organic farming methods believe that the increased land needed to farm organic food could potentially destroy the rainforests and wipe out many ecosystems.

Nutrient Leaching

According to the meta-analysis of 71 studies, nitrogen leaching, nitrous oxide emissions, ammonia emissions, eutrophication potential and acidification potential were higher for organic products, although in one study "nitrate leaching was 4.4-5.6 times higher in conventional plots than organic plots".

Excess nutrients in lakes, rivers, and groundwater can cause algal blooms, eutrophication, and subsequent dead zones. In addition, nitrates are harmful to aquatic organisms by themselves.

Land Use

The Oxford meta-analysis of 71 studies proved that organic farming requires 84% more land, mainly due to lack of nutrients but sometimes due to weeds, diseases or pests, lower yielding animals and land required for fertility building crops. While organic farming does not necessarily save land for wildlife habitats and forestry in all cases, the most modern breakthroughs in organic are addressing these issues with success.

Professor Wolfgang Branscheid says that organic animal production is not good for the environment, because organic chicken requires doubly as much land as "conventional" chicken and organic pork a quarter more. According to a calculation by Hudson Institute, organic beef requires triply as much land. On the other hand, certain organic methods of animal husbandry have been shown to restore desertified, marginal, and/or otherwise unavailable land to agricultural productivity and wildlife. Or by getting both forage and cash crop production from the same fields simultaneously, reduce net land use.

In England organic farming yields 55% of normal yields. While in other regions of the world, organic methods have started producing record yields.

Pesticides

A sign outside of an organic apple orchard in Pateros, Washington reminding orchardists not to spray pesticides on these trees.

Food Quality and Safety

While there may be some differences in the amounts of nutrients and anti-nutrients when organically produced food and conventionally produced food are compared, the variable nature of food production and handling makes it difficult to generalize results, and there is insufficient evidence to make claims that organic food is safer or healthier than conventional food. Claims that organic food tastes better are not supported by evidence.

Soil Conservation

Supporters claim that organically managed soil has a higher quality and higher water retention. This may help increase yields for organic farms in drought years. Organic farming can build up soil organic matter better than conventional no-till farming, which suggests long-term yield benefits from organic farming. An 18-year study of organic methods on nutrient-depleted soil concluded that conventional methods were superior for soil fertility and yield for nutrient-depleted soils in cold-temperate climates, arguing that much of the benefit from organic farming derives from imported materials that could not be regarded as self-sustaining.

In *Dirt: The Erosion of Civilizations*, geomorphologist David Montgomery outlines a coming crisis from soil erosion. Agriculture relies on roughly one meter of topsoil, and that is being depleted ten times faster than it is being replaced. No-till farming, which some claim depends upon pesticides, is one way to minimize erosion. However, a 2007 study by the USDA's Agricultural Research Service has found that manure applications in tilled organic farming are better at building up the soil than no-till.

Biodiversity

The conservation of natural resources and biodiversity is a core principle of organic production. Three broad management practices (prohibition/reduced use of chemical pesticides and inorganic fertilizers; sympathetic management of non-cropped habitats; and preservation of mixed farming) that are largely intrinsic (but not exclusive) to organic farming are particularly beneficial for farmland wildlife. Using practices that attract or introduce beneficial insects, provide habitat for birds and mammals, and provide conditions that increase soil biotic diversity serve to supply vital ecological services to organic production systems. Advantages to certified organic operations that implement these types of production practices include: 1) decreased dependence on outside fertility inputs; 2) reduced pest management costs; 3) more reliable sources of clean water; and 4) better pollination.

Nearly all non-crop, naturally occurring species observed in comparative farm land practice studies show a preference for organic farming both by abundance and diversity. An average of 30% more species inhabit organic farms. Birds, butterflies, soil microbes, beetles, earthworms, spiders, vegetation, and mammals are particularly affected. Lack of herbicides and pesticides improve biodiversity fitness and population density. Many weed species attract beneficial insects that improve soil qualities and forage on weed pests. Soil-bound organisms often benefit because of increased bacteria populations due to natural fertilizer such as manure, while experiencing reduced intake of herbicides and pesticides. Increased biodiversity, especially from beneficial soil microbes and mycorrhizae have been proposed as an explanation for the high yields experienced by some organic plots, especially in light of the differences seen in a 21-year comparison of organic and control fields.

Biodiversity from organic farming provides capital to humans. Species found in organic farms enhance sustainability by reducing human input (e.g., fertilizers, pesticides).

The USDA's Agricultural Marketing Service (AMS) published a *Federal Register* notice on 15 January 2016, announcing the National Organic Program (NOP) final guidance on Natural Resources and Biodiversity Conservation for Certified Organic Operations. Given the broad scope of natural resources which includes soil, water, wetland, woodland and wildlife, the guidance provides examples of practices that support the underlying conservation principles and demonstrate compliance with USDA organic regulations § 205.200. The final guidance provides organic certifiers and farms with examples of production practices that support conservation principles and comply with the USDA organic regulations, which require operations to maintain or improve natural resources. The final guidance also clarifies the role of certified operations (to submit an OSP to a certifier), certifiers (ensure that the OSP describes or lists practices that explain the operator's monitoring plan and practices to support natural resources and biodiversity conservation), and inspectors (onsite inspection) in the implementation and verification of these production practices.

A wide range of organisms benefit from organic farming, but it is unclear whether organic methods confer greater benefits than conventional integrated agri-environmental programs. Organic farming is often presented as a more biodiversity-friendly practice, but the generality of the beneficial effects of organic farming is debated as the effects appear often species- and context-dependent, and current research has highlighted the need to quantify the relative effects of local- and landscape-scale management on farmland biodiversity. There are four key issues when comparing the impacts on biodiversity of organic and conventional farming: (1) It remains unclear whether a holistic whole-farm approach (i.e. organic) provides greater benefits to bio-diversity than carefully targeted prescriptions applied to relatively small areas of cropped and/or non-cropped habitats within conventional agriculture (i.e. agri-environment schemes); (2) Many comparative studies encounter methodological problems, limiting their ability to draw quantitative conclusions; (3) Our knowledge of the impacts of organic farming in pastoral and upland agriculture is limited; (4) There remains a pressing need for longitudinal, system-level studies in order to address these issues and to fill in the gaps in our knowledge of the impacts of organic farming, before a full appraisal of its potential role in biodiversity conservation in agroecosystems can be made.

Regional Support for Organic Farming

India

In India, states such as Sikkim and Kerala have planned to shift to fully organic cultivation by 2015 and 2016 respectively.

Development of Organic Farming Over the Years

History of Organic Farming

Traditional farming (of many particular kinds in different eras and places) was the original type of agriculture, and has been practiced for thousands of years. All traditional farming is now considered to be "organic farming" although at the time there were no known inorganic methods. For example, forest gardening, a fully organic food production system which dates from prehistoric times, is thought to be the world's oldest and most resilient agroecosystem. After the industrial revolution had introduced inorganic methods, most of which were not well developed and had serious side effects, an organic movement began in the 1940s as a reaction to agriculture's growing reliance on synthetic fertilizers and pesticides. The history of this modern revival of organic farming dates back to the first half of the 20th century at a time when there was a growing reliance on these new synthetic, inorganic methods.

Pre-World War II

The first 40 years of the 20th century saw simultaneous advances in biochemistry and engineering that rapidly and profoundly changed farming. The introduction of the gasoline-powered internal combustion engine ushered in the era of the tractor and made possible hundreds of mechanized farm implements. Research in plant breeding led to the commercialization of hybrid seed. And a new manufacturing process made nitrogen fertilizer — first synthesized in the mid-19th century — affordably abundant. These factors changed the labor equation: there were almost no tractors in the US around 1910, but over 3,000,000 by 1950; in 1900, it took one farmer to feed 2.5 people, but currently the ratio is 1 to well over 100. Fields grew bigger and cropping more specialized to make more efficient use of machinery. The reduced need for manual labour and animal labour that machinery, herbicides, and fertilizers made possible created an era in which the mechanization of agriculture evolved rapidly.

Consciously organic agriculture (as opposed to traditional agricultural methods from before the inorganic options existed, which always employed only organic means) began more or less simultaneously in Central Europe and India. The British botanist Sir Albert Howard is often referred to as the father of modern organic agriculture, because he was the first to apply modern scientific knowledge and methods to traditional agriculture. From 1905 to 1924, he and his wife Gabrielle, herself a plant physiologist, worked as agricultural advisers in Pusa, Bengal, where they documented traditional Indian farming practices and came to regard them as superior to their conventional agriculture science. Their research and further development of these methods is recorded in his writings, notably, his 1940 book, *An Agricultural Testament*, which influenced many scientists and farmers of the day.

In Germany, Rudolf Steiner's development, biodynamic agriculture, was probably the first comprehensive system of what we now call organic farming. This began with a lecture series Steiner presented at a farm in Koberwitz (Kobierzyce now in Poland) in 1924. Steiner emphasized the farmer's role in guiding and balancing the interaction of the animals, plants and soil. Healthy animals depended upon healthy plants (for their food), healthy plants upon healthy soil, healthy soil upon healthy animals (for the manure). His system was based on his philosophy of anthroposophy rather than a good understanding of science. To develop his system of farming, Steiner established an international research group called the Agricultural Experimental Circle of Anthroposophical Farmers and Gardeners of the General Anthroposophical Society.

In 1909, American agronomist F.H. King toured China, Korea, and Japan, studying traditional fertilization, tillage, and general farming practices. He published his findings in *Farmers of Forty Centuries* (1911, Courier Dover Publications, ISBN 0-486-43609-8). King foresaw a "world movement for the introduction of new and improved methods" of agriculture and in later years his book became an important organic reference.

The term *organic farming* was coined by Lord Northbourne in his book *Look to the Land* (written in 1939, published 1940). From his conception of "the farm as organism," he described a holistic, ecologically balanced approach to farming.

In 1939 Lady Eve Balfour launched the Haughley Experiment on farmland in England. Lady Balfour believed that mankind's health and future depended on how the soil was used, and that non-intensive farming could produce more wholesome food. The experiment was run to generate data in support of these beliefs. Four years later, she published *The Living Soil*, based on the initial findings of the Haughley Experiment. Widely read, it led to the formation of a key international organic advocacy group, the Soil Association.

In Japan, Masanobu Fukuoka, a microbiologist working in soil science and plant pathology, began to doubt the modern agricultural movement. In 1937, he quit his job as a research scientist, returned to his family's farm in 1938, and devoted the next 60 years to developing a radical no-till organic method for growing grain and many other crops, now known as natural farming (自然農法 *shizen nōhō?*), nature farming, 'do–nothing' farming or Fukuoka farming.

Post-World War II

Technological advances during World War II accelerated post-war innovation in all aspects of agriculture, resulting in large advances in mechanization (including large-scale irrigation), fertilization, and pesticides. In particular, two chemicals that had been produced in quantity for warfare, were repurposed for peace-time agricultural uses. Ammonium nitrate, used in munitions, became an abundantly cheap source of nitrogen. And a range of new pesticides appeared: DDT, which had been used to control disease-carrying insects around troops, became a general insecticide, launching the era of widespread pesticide use.

At the same time, increasingly powerful and sophisticated farm machinery allowed a single farmer to work larger areas of land and fields grew bigger.

In 1944, an international campaign called the Green Revolution was launched in Mexico with private funding from the US. It encouraged the development of hybrid plants, chemical controls, large-scale irrigation, and heavy mechanization in agriculture around the world.

During the 1950s, sustainable agriculture was a topic of scientific interest, but research tended to concentrate on developing the new chemical approaches. One of the reasons for this, which informed and guided the ongoing Green Revolution, was the widespread belief that high global population growth, which was demonstrably occurring, would soon create worldwide food shortages unless humankind could rescue itself through ever higher agricultural technology. At the same time, however, the adverse effects of "modern" farming continued to kindle a small but growing organic movement. For example, in the US, J.I. Rodale began to popularize the term and methods of organic growing, particularly to consumers through promotion of organic gardening.

In 1962, Rachel Carson, a prominent scientist and naturalist, published *Silent Spring*, chronicling the effects of DDT and other pesticides on the environment. A bestseller in many countries, including the US, and widely read around the world, *Silent Spring* is widely considered as being a key factor in the US government's 1972 banning of DDT. The book and its author are often credited with launching the worldwide environmental movement.

In the 1970s, global movements concerned with pollution and the environment increased their focus on organic farming. As the distinction between organic and conventional food became clearer, one goal of the organic movement was to encourage consumption of locally grown food, which was promoted through slogans like "Know Your Farmer, Know Your Food".

In 1972, the International Federation of Organic Agriculture Movements (IFOAM) was founded in Versailles, France and dedicated to the diffusion and exchange of information on the principles and practices of organic agriculture of all schools and across national and linguistic boundaries.

In 1975, Fukuoka released his book, *The One-Straw Revolution*, with a strong impact in certain areas of the agricultural world. His approach to small-scale grain production emphasized a meticulous balance of the local farming ecosystem, and a minimum of human interference and labor.

In the U.S. during the 1970s and 1980s, J.I. Rodale and his Rodale Press (now Rodale, Inc.) led the way in getting Americans to think about the side effects of nonorganic methods, and the advantages of organic ones. The press's books offered how-to information and advice to Americans interested in trying organic gardening and farming.

In 1984, Oregon Tilth established an early organic certification service in the United States.

In the 1980s, around the world, farming and consumer groups began seriously pressuring for government regulation of organic production. This led to legislation and certification standards being enacted through the 1990s and to date. In the United States, the Organic Foods Production Act of 1990 tasked the USDA with developing national standards for organic products, and the final rule establishing the National Organic Program was first published in the Federal Register in 2000

In Havana, Cuba, the loss of Soviet economic support following the collapse of the Soviet Union in 1991 led to a focus on local agricultural production and the development of a unique state-supported urban organic agriculture program called organopónicos.

Since the early 1990s, the retail market for organic farming in developed economies has been growing by about 20% annually due to increasing consumer demand. Concern for the quality and safety of food, and the potential for environmental damage from conventional agriculture, are apparently responsible for this trend.

Twenty-First Century

Throughout this history, the focus of agricultural research and the majority of publicized scientific findings has been on chemical, not organic, farming. This emphasis has continued to biotechnologies like genetic engineering. One recent survey of the UK's leading government funding agency for bioscience research and training indicated 26 GM crop projects, and only one related to organic agriculture. This imbalance is largely driven by agribusiness in general, which, through research funding and government lobbying, continues to have a predominating effect on agriculture-related science and policy.

Agribusiness is also changing the rules of the organic market. The rise of organic farming was driven by small, independent producers and by consumers. In recent years, explosive organic market growth has encouraged the participation of agribusiness interests. As the volume and variety of "organic" products increases, the viability of the small-scale organic farm is at risk, and the meaning of organic farming as an agricultural method is ever more easily confused with the related but separate areas of organic food and organic certification.

Forest Farming

Forest farming is the cultivation of high-value specialty crops under a forest canopy that is intentionally modified or maintained to provide shade levels and habitat that favor growth and enhance production levels. Forest farming encompasses a range of cultivated systems from introducing plants into the understory of a timber stand to modifying forest stands to enhance the marketability and sustainable production of existing plants.

Forest farming is a type of agroforestry practice characterized by the "four I's": intentional, integrated, intensive and interactive. Agroforestry is a land management system that combines trees with crops or livestock, or both, on the same piece of land. It focuses on increasing benefits to the landowner as well as maintaining forest integrity and environmental health. The practice involves cultivating non-timber forest products or niche crops, some of which, such as ginseng or shiitake mushrooms, can have high market value.

Non-timber forest products (NTFPs) are plants, parts of plants, fungi, and other biological materials harvested from within and on the edges of natural, manipulated, or disturbed forests. Examples of crops are ginseng, shiitake mushrooms, decorative ferns, and pine straw. Products typically fit into the following categories: edible, medicinal and dietary supplements, floral or decorative, or specialty wood-based products.

History

Toyohiko Kagawa, forest farming pioneer.

Forest farming, though not always by that name, is practiced around the world. For centuries, humans have relied on fruits, nuts, seeds, parts of foliage and pods from trees and shrubs in the forests to feed themselves and their livestock. Over time, certain species have been selected for cultivation near homes or livestock to provide food or medicine. For example, in the southern United States, mulberry trees are used as a feedstock for pigs and often cultivated near pig quarters.

In 1929, J. Russell Smith, Emeritus Professor of Economic Geography at Columbia University, published "Tree Crops – A Permanent Agriculture" which stated that crop-yielding trees could provide useful substitutes for cereals in animal feeding programs, as well as conserve environmental health. Toyohiko Kagawa read and was heavily influenced by Smith's publication and began experimental cultivation under trees in Japan during the 1930s. Through forest farming, or three-dimensional forestry, Kagawa addressed problems of soil erosion by persuading many of Japan's upland farmers to plant fodder trees to conserve soil, supply food and feed animals. He combined extensive plantings of walnut trees, harvested the nuts and fed them to the pigs, then sold the pigs as a source of income. When the walnut trees matured, they were sold for timber and more trees were planted so that there was a continuous cycle of economic cropping that provided both short-term and long-term income to the small landowner. The success of these trials prompted similar research in other countries. Unfortunately, World War II disrupted communication and slowed advances in forest farming. In the mid-1950s research resumed in places such as southern Africa. Kagawa was also an inspiration to Robert Hart pioneered forest gardening in temperate climates in the sixties in Shropshire, England.

In earlier years, livestock were often considered part of the forest farming system. Now they are typically excluded and agroforestry systems that integrate trees, forages and livestock are referred to as silvopastures. Because forest farming combines ecological stability of natural forests and productive agriculture systems, it is considered to have great potential for regenerating soils, restoring ground water supplies, controlling floods and droughts and cultivating marginal lands. In addition to these benefits for re-establishing productive forests on marginal lands, forest farming is way to add financial value while conserving land that is currently forested, as discussed in the methods section.

In more recent years, there has been growing interest in locally grown and organic foods throughout the United States. There has been an increase in farmer's markets and community-supported agriculture small enterprises. These have also become outlets for NTFPs. In order to stay competitive, many farmers look to add unique crops to their product line. With the quantity and quality of resources developing online that offer tutorials and educational information on how to create and maintain forest farms, forest gardens, how to cultivate specific crops such as shiitake mushrooms and how to successfully market these items, forest farming is expanding as a viable land management practice. Good places to look for research-based resources are the USDA National Agroforestry Center's publication section, the Center for Agroforestry at the University of Missouri, the Cornell Cooperative Extension, the Non-Timber Forest Products website by The Virginia Tech Department of Wood Science and Forest Products, the USDA Forest Service Southern Research Station and the Top of the Ozarks RC&D in Missouri and the collaborative Forest Farming community of practice on eXtension.org, the online presence of the Cooperative Extension System of the US Land Grant Universities.

Principles

Forest farming principles constitute an ecological approach to forest management. Forest resources are judiciously used while biodiversity and wildlife habitat are conserved. Forest farms have the potential to restore ecological balance to fragmented second growth forests through intentional manipulation to create the desired forest ecosystem.

In some instances, the intentional introduction of species for botanicals, medicinals, food or decorative products is accomplished using existing forests. The tree cover, soil type, water supply, land form and other site characteristics determine what species will thrive. Developing an understanding of species/site relationships as well as understanding the site limitations is necessary to utilize these resources for production needs, while conserving adequate resources for the long-term health of the forest.

Apart from the environmental benefits, forest farming can increase the economic value of forest property and provide short- and long-term benefits to the landowner. Forest farming provides economic return from intact forest ecosystems, but timber sales can remain part of the long-term management strategy.

Methods

Forest farming methods may include: Intensive, yet careful thinning of overstocked, suppressed tree stands; multiple integrated entries to accomplish thinning so that systemic shock is minimized; and interactive management to maintain a cross-section of healthy trees and shrubs of all ages and species. Physical disturbance to the surrounding area should be minimized. The following are forest farming techniques described in the Training Manual produced by the Center for Agroforestry at the University of Missouri.

Level of Management Required

1. Forest gardening is the most intensive of forest farming methods. In addition to thinning the overstory, this method involves clearing the understory of undesirable vegetation and other practices that are closely related to agronomy (tillage, fertilization, weeding, and control of disease and insects and wildlife management). Due to input levels, this method often produces lower valued products compared to other methods. Forest gardens take advantage of the vertical levels of light availability and space under the forest canopy so that more than one crop can be grown at once if desired.

2. Wild-simulated seeks to maintain a natural growing environment, yet enriches local NTFP populations to create an abundant renewable supply of the products. Minimal disturbance and natural growing conditions ensure products will be similar in appearance and quality of those harvested from the wild. Rather than till, practitioners often rake leaves to expose soil, sow seed directly onto the ground, and then cover with leaves again. Since this method produces NTFPs that closely resemble wild plants; they often command a higher price than NTFPs produced using the forest gardening method.

3. Forest tending involves adjusting tree crown density to manipulate light levels that favor natural reproduction of desirable NTFPs. This low intensity management approach does not involve supplemental planting to increase populations of desired NTFPs.

4. Wildcrafting is the harvesting of naturally growing NTFPs. It is not considered a forest farming practice since there is no human involvement in the plant's establishment and maintenance. However, wildcrafters often take steps to protect NTFPs with future harvests in mind. It becomes agroforestry once forest thinnings, or other inputs, are applied to sustain or maintain plant populations that might otherwise succumb to successional changes in the forest. The most important difference between forest farming and wildcrafting is that forest farming intentionally produces NTFPS, whereas wildcrafting seeks and gathers from naturally growing NTFPs.

Production Considerations

Forest farming can be a small business opportunity for landowners and requires careful planning, including a business and marketing plan. Learning how to market the NTFPs on the Internet is an option, but may entail higher shipping costs. Landowners should consider all options for selling their products including, farmer's markets or restaurants that focus on locally grown ingredients. The development phase should include a forest management plan that states the landowner's objectives and a resource inventory. Start-up costs should be analyzed as specific equipment may be necessary to harvest or process the product, whereas other crops require minimal initial investment. Local incentives for sustainable forest management, as well as regulations and policies should be explored. The Convention on International Trade in Endangered Species of Wild Fauna and Flora (CITES) regulates international trade of certain plant (American ginseng and goldenseal) and animal species. To be legally exported, regulated plants must be harvested and records kept according to CITES rules and restrictions. Many states also have harvesting regulations for certain native plants that are searchable online. Another good source to start with on information is the Medicinal Plants at Risk 2008 report, by the Center for Biological Diversity] in the U.S.

Examples of Crops

Medicinal Herbs:

- Ginseng (*Panax quinquefolius*)
- Black Cohosh (*Actaea racemosa*)
- Goldenseal (*Hydrastis canadensis*)
- Bloodroot (*Sanguinaria canadensis*)
- Pacific yew (*Taxus brevifolia*)
- Mayapple (*Podophyllum peltatum*)
- Saw palmetto (*Serenoa repens*)
- American Pokeweed (*Phytolacca americana*)

Nuts:

- Black walnut (*Juglans nigra*)
- Hazelnut (*Corylus avellana*)
- Shagbark hickory (*Carya ovata*)
- Beechnut (*Fagus sylvatica*)

Fruit:

- Pawpaw (*Asimina triloba*)
- Currants (*Ribes spp*)
- Elderberry (*Sambucus spp*)
- Serviceberry (*Amelanchier spp*)
- Blackberry (*Rubus spp*)

- Huckleberry (*Gaylussacia brachycera*)

Other Food Crops:

- Ramps (wild leeks) (*Allium tricoccum*)
- Syrups (maple)
- Honey
- Mushrooms
- Other edible roots

Other Products: (Mulch, Decoratives, Crafts, Dyes)

- Pine straw
- Willow twigs
- Vines
- Beargrass (*Xerophyllum tenax*)
- Ferns
- Pine cones
- Moss

Native Ornamentals:

- Rhododendron (*Rhododendron catawbiense*)
- Highbush cranberry (*Viburnum trilobum*)
- Flowering dogwood (*Cornus florida*)

Extensive Farming

Extensive farming or extensive agriculture (as opposed to intensive farming) is an agricultural production system that uses small inputs of labor, fertilizers, and capital, relative to the land area being farmed.

Extensive farming most commonly refers to sheep and cattle farming in areas with low agricultural productivity, but can also refer to large-scale growing of wheat, barley, cooking oils and other grain crops in areas like the Murray-Darling Basin. Here, owing to the extreme age and poverty of the soils, yields per hectare are very low, but the flat terrain and very large farm sizes mean yields per unit of labour are high. Nomadic herding is an extreme example of extensive farming, where herders move their animals to use feed from occasional rainfalls.

A small farm in the Swiss mountains. The land here is mostly rock and the slopes are very steep – likely unusable for agriculture, but can provide productive conditions for pigs

Geography

Extensive farming is found in the mid-latitude sections of most continents, as well as in desert regions where water for cropping is not available. The nature of extensive farming means it requires less rainfall than intensive farming. The farm is usually large in comparison with the numbers working and money spent on it. In most parts of Western Australia, pastures are so poor that only one sheep to the square mile can be supported

Just as the demand has led to the basic division of cropping and pastoral activities, these areas can also be subdivided depending on the region's rainfall, vegetation type and agricultural activity within the area and the many other parentheses related to this data.

Advantages

Extensive farming has a number of advantages over intensive farming:

1. Less labour per unit areas is required to farm large areas, especially since expensive alterations to land (like terracing) are completely absent.

2. Mechanisation can be used more effectively over large, flat areas.

3. Greater efficiency of labour means generally lower product prices.

4. Animal welfare is generally improved because animals are not kept in stifling conditions.

5. Lower requirements of inputs such as fertilizers.

6. If animals are grazed on pastures native to the locality, there is less likely to be problems with exotic species.

7. Local environment and soil are not damaged by overuse of chemicals.

8. The use of machinery and scientific methods of farming produce a large quantity of crops.

Disadvantages

Extensive farming can have the following problems:

1. Yields tend to be much lower than with intensive farming in the short term.

2. Large land requirements limit the habitat of wild species (in some cases, even very low stocking rates can be dangerous), as is the case with intensive farming

Intensive Farming

Intensive farming or intensive agriculture also known as industrial agriculture is characterized by a low fallow ratio and higher use of inputs such as capital and labour per unit land area. This is in contrast to traditional agriculture in which the inputs per unit land are lower.

Intensive animal husbandry involves either large numbers of animals raised on limited land, usually confined animal feeding operations (CAFO) often referred to as factory farms, or managed intensive rotational grazing (MIRG). Both increase the yields of food and fiber per acre as compared to traditional animal husbandry. In a CAFO feed is brought to the animals, which are seldom moved, while in MIRG the animals are repeatedly moved to fresh forage.

Intensive crop agriculture is characterised by innovations designed to increase yield. Techniques include planting multiple crops per year, reducing the frequency of fallow years and improving cultivars. It also involves increased use of fertilizers, plant growth regulators, pesticides and mechanization, controlled by increased and more detailed analysis of growing conditions, including weather, soil, water, weeds and pests.

This system is supported by ongoing innovation in agricultural machinery and farming methods, genetic technology, techniques for achieving economies of scale, logistics and data collection and analysis technology. Intensive farms are widespread in developed nations and increasingly prevalent worldwide. Most of the meat, dairy, eggs, fruits and vegetables available in supermarkets are produced by such farms.

Smaller intensive farms usually include higher inputs of labor and more often use sustainable intensive methods. The farming practices commonly found on such farms are referred to as appropriate technology. These farms are less widespread in both developed countries and worldwide, but are growing more rapidly. Most of the food available in specialty markets such as farmers markets is produced by these smallholder farms.

History

Early 20th-century image of a tractor ploughing an alfalfa field

Agricultural development in Britain between the 16th century and the mid-19th century saw a massive increase in agricultural productivity and net output. This in turn supported unprecedented population growth, freeing up a significant percentage of the workforce, and thereby helped enable the Industrial Revolution. Historians cited enclosure, mechanization, four-field crop rotation, and selective breeding as the most important innovations.

Industrial agriculture arose along with the Industrial Revolution. By the early 19th century, agricultural techniques, implements, seed stocks and cultivars had so improved that yield per land unit was many times that seen in the Middle Ages.

The industrialization phase involved a continuing process of mechanization. Horse-drawn machinery such as the McCormick reaper revolutionized harvesting, while inventions such as the cotton gin reduced the cost of processing. During this same period, farmers began to use steam-powered threshers and tractors, although they were expensive and dangerous. In 1892, the first gasoline-powered tractor was successfully developed, and in 1923, the International Harvester Farmall tractor became the first all-purpose tractor, marking an inflection point in the replacement of draft animals with machines. Mechanical harvesters (combines), planters, transplanters and other equipment were then developed, further revolutionizing agriculture. These inventions increased yields and allowed individual farmers to manage increasingly large farms.

The identification of nitrogen, potassium, and phosphorus (NPK) as critical factors in plant growth led to the manufacture of synthetic fertilizers, further increasing crop yields. In 1909 the Haber-Bosch method to synthesize ammonium nitrate was first demonstrated. NPK fertilizers stimulated the first concerns about industrial agriculture, due to concerns that they came with serious side effects such as soil compaction, soil erosion and declines in overall soil fertility, along with health concerns about toxic chemicals entering the food supply.

The identification of carbon as a critical factor in plant growth and soil health, particularly in the form of humus, led to so-called *sustainable agriculture*, alternative forms of intensive agriculture that also surpass traditional agriculture, without side effects or health issues. Farmers adopting this approach were initially referred to as *humus farmers*, later as *organic farmers*.

The discovery of vitamins and their role in nutrition, in the first two decades of the 20th century, led to vitamin supplements, which in the 1920s allowed some livestock to be raised indoors, reducing their exposure to adverse natural elements. Chemicals developed for use in World War II gave rise to synthetic pesticides.

Following World War II, synthetic fertilizer use increased rapidly, while sustainable intensive farming advanced much more slowly. Most of the resources in developed nations went to improving industrial intensive farming, and very little went to improving organic farming. Thus, particularly in the developed nations, industrial intensive farming grew to become the dominant form of agriculture.

The discovery of antibiotics and vaccines facilitated raising livestock in CAFOs by reducing diseases caused by crowding. Developments in logistics and refrigeration as well as processing technology made long-distance distribution feasible.

Between 1700 and 1980, "the total area of cultivated land worldwide increased 466%" and yields increased dramatically, particularly because of selectively bred high-yielding varieties, fertilizers, pesticides, irrigation and machinery. Global agricultural production doubled between 1820 1920; between 1920 and 1950; between 1950 and 1965; and again between 1965 and 1975 to feed a global population that grew from one billion in 1800 to 6.5 billion in 2002. The number of people involved in farming in industrial countries dropped, from 24 percent of the American population to 1.5 percent in 2002. In 1940, each farmworker supplied 11 consumers, whereas in 2002, each worker supplied 90 consumers. The number of farms also decreased and their ownership became more concentrated. In 2000 in the U.S., four companies produced 81 percent of cows, 73 percent of sheep, 57 percent of pigs, and 50 percent of chickens, cited as an example of "vertical integration" by the president of the U.S. National Farmers' Union. Between 1967 and 2002 the one million pig farms in America consolidated into 114,000 with 80 million pigs (out of 95 million) produced each year on factory farms, according to the U.S. National Pork Producers Council. According to the Worldwatch Institute, 74 percent of the world's poultry, 43 percent of beef, and 68 percent of eggs are produced this way.

Concerns over the sustainability of industrial agriculture, which has become associated with decreased soil quality, and over the environmental effects of fertilizers and pesticides, have not subsided. Alternatives such as integrated pest management (IPM) have had little impact because policies encourage the use of pesticides and IPM is knowledge-intensive. These concerns sustained the organic movement and caused a resurgence in sustainable intensive farming and funding for the development of appropriate technology.

Famines continued throughout the 20th century. Through the effects of climactic events, government policy, war and crop failure, millions of people died in each of at least ten famines between the 1920s and the 1990s.

Techniques and Technologies

Livestock

A commercial chicken house raising broiler pullets for meat.

Confined Animal Feeding Operations

Intensive livestock farming, also called "factory farming" is a term referring to the process of raising livestock in confinement at high stocking density. "Concentrated animal feeding operations" (CAFO) or "intensive livestock operations", can hold large numbers (some up to hundreds of thousands) of cows, hogs, turkeys or chickens, often indoors. The essence of such farms is the concentration of livestock in a given space. The aim is to provide maximum output at the lowest possible cost and with the greatest level of food safety. The term is often used pejoratively. However, CAFOs have dramatically increased the production of food from animal husbandry worldwide, both in terms of total food produced and efficiency.

Food and water is delivered to the animals, and therapeutic use of antimicrobial agents, vitamin supplements and growth hormones are often employed. Growth hormones are not used on chickens nor on any animal in the European Union. Undesirable behaviours often related to the stress of confinement led to a search for docile breeds (e.g., with natural dominance behaviours bred out), physical restraints to stop interaction, such as individual cages for chickens, or physically modification such as the de-beaking of chickens to reduce the harm of fighting.

The CAFO designation resulted from the 1972 US Federal Clean Water Act, which was enacted to protect and restore lakes and rivers to a "fishable, swimmable" quality. The United States Environmental Protection Agency (EPA) identified certain animal feeding operations, along with many other types of industry, as "point source" groundwater polluters. These operations were subjected to regulation.

In 17 states in the U.S., isolated cases of groundwater contamination were linked to CAFOs. For example, the ten million hogs in North Carolina generate 19 million tons of waste per year. The U.S. federal government acknowledges the waste disposal issue and requires that animal waste be stored in lagoons. These lagoons can be as large as 7.5 acres (30,000 m²). Lagoons not protected with an impermeable liner can leak into groundwater under some conditions, as can runoff from manure used as fertilizer. A lagoon that burst in 1995 released 25 million gallons of nitrous sludge in North Carolina's New River. The spill allegedly killed eight to ten million fish.

The large concentration of animals, animal waste and dead animals in a small space poses ethical issues to some consumers. Animal rights and animal welfare activists have charged that intensive animal rearing is cruel to animals.

Other concerns include persistent noxious odor, the effects on human health and the role of antibiotics use in the rise of resistant infectious bacteria.

According to the U.S. Centers for Disease Control and Prevention (CDC), farms on which animals are intensively reared can cause adverse health reactions in farm workers. Workers may develop acute and/or chronic lung disease, musculoskeletal injuries and may catch (zoonotic) infections from the animals.

Managed Intensive Rotational Grazing

Managed Intensive Rotational Grazing (MIRG), also known as cell grazing, mob grazing and holistic managed planned grazing, is a variety of forage use in which herds/flocks are regularly and systematically moved to fresh, rested grazing areas to maximize the quality and quantity of forage growth. MIRG can be used with cattle, sheep, goats, pigs, chickens, turkeys, ducks and other animals. The herds graze one portion of pasture, or a paddock, while allowing the others to recover. Resting grazed lands allows the vegetation to renew energy reserves, rebuild shoot systems, and deepen root systems, resulting in long-term maximum biomass production. MIRG is especially effective because grazers thrive on the more tender younger plant stems. MIRG also leave parasites behind to die off minimizing or eliminating the need for de-wormers. Pasture systems alone can allow grazers to meet their energy requirements, and with the increased productivity of MIRG systems, the animals obtain the majority of their nutritional needs, in some cases all, without the supplemental feed sources that are required in continuous grazing systems or CAFOs.

Crops

The Green Revolution transformed farming in many developing countries. It spread technologies that had already existed, but had not been widely used outside of industrialized nations. These technologies included "miracle seeds", pesticides, irrigation and synthetic nitrogen fertilizer.

Seeds

In the 1970s scientists created strains of maize, wheat, and rice that are generally referred to as high-yielding varieties (HYV). HYVs have an increased nitrogen-absorbing potential compared to other varieties. Since cereals that absorbed extra nitrogen would typically lodge (fall over) before harvest, semi-dwarfing genes were bred into their genomes. Norin 10 wheat, a variety developed by Orville Vogel from Japanese dwarf wheat varieties, was instrumental in developing wheat cultivars. IR8, the first widely implemented HYV rice to be developed by the International Rice Research Institute, was created through a cross between an Indonesian variety named "Peta" and a Chinese variety named "Dee Geo Woo Gen."

With the availability of molecular genetics in Arabidopsis and rice the mutant genes responsible (*reduced height (rht), gibberellin insensitive (gai1)* and *slender rice (slr1)*) have been cloned and identified as cellular signalling components of gibberellic acid, a phytohormone involved in regulating stem growth via its effect on cell division. Photosynthetic investment in the stem is reduced dramatically as the shorter plants are inherently more mechanically stable. Nutrients become redirected to grain production, amplifying in particular the yield effect of chemical fertilisers.

HYVs significantly outperform traditional varieties in the presence of adequate irrigation, pesticides and fertilizers. In the absence of these inputs, traditional varieties may outperform HYVs. They were developed as F1 hybrids, meaning seeds need to be purchased every season to obtain maximum benefit, thus increasing costs.

Crop rotation

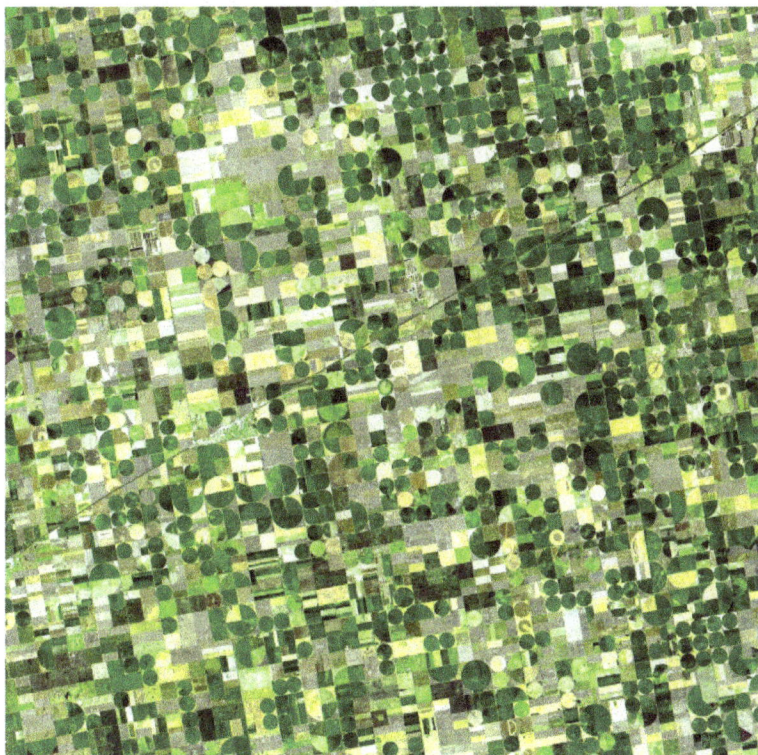

Satellite image of circular crop fields in Haskell County, Kansas in late June 2001. Healthy, growing crops of corn and sorghum are green (Sorghum may be slightly paler). Wheat is brilliant gold. Fields of brown have been recently harvested and plowed under or have lain in fallow for the year.

Crop rotation or crop sequencing is the practice of growing a series of dissimilar types of crops in the same space in sequential seasons for benefits such as avoiding pathogen and pest buildup that occurs when one species is continuously cropped. Crop rotation also seeks to balance the nutrient demands of various crops to avoid soil nutrient depletion. A traditional component of crop rotation is the replenishment of nitrogen through the use of legumes and green manure in sequence with cereals and other crops. Crop rotation can also improve soil structure and fertility by alternating deep-rooted and shallow-rooted plants. One technique is to plant multi-species cover crops between commercial crops. This combines the advantages of intensive farming with continuous cover and polyculture.

Irrigation

Overhead irrigation, center pivot designed

Crop irrigation accounts for 70% of the world's fresh water use.

Flood irrigation, the oldest and most common type, is typically unevenly distributed, as parts of a field may receive excess water in order to deliver sufficient quantities to other parts. Overhead irrigation, using center-pivot or lateral-moving sprinklers, gives a much more equal and controlled distribution pattern. Drip irrigation is the most expensive and least-used type, but delivers water to plant roots with minimal losses.

Water catchment management measures include recharge pits, which capture rainwater and run-off and use it to recharge groundwater supplies. This helps in the replenishment of groundwater wells and eventually reduces soil erosion. Dammed rivers creating Reservoirs store water for irrigation and other uses over large areas. Smaller areas sometimes use irrigation ponds or ground-water.

Weed Control

In agriculture, systematic weed management is usually required, often performed by machines such as cultivators or liquid herbicide sprayers. Herbicides kill specific targets while leaving the crop relatively unharmed. Some of these act by interfering with the growth of the weed and are often based on plant hormones. Weed control through herbicide is made more difficult when the weeds become resistant to the herbicide. Solutions include:

- Cover crops (especially those with allelopathic properties) that out-compete weeds or in-hibit their regeneration.

- Multiple herbicides, in combination or in rotation

- Strains genetically engineered for herbicide tolerance

- Locally adapted strains that tolerate or out-compete weeds

- Tilling

- Ground cover such as mulch or plastic

- Manual removal

- Mowing

- Grazing

- Burning

Terracing

In agriculture, a terrace is a leveled section of a hilly cultivated area, designed as a method of soil conservation to slow or prevent the rapid surface runoff of irrigation water. Often such land is formed into multiple terraces, giving a stepped appearance. The human landscapes of rice cultivation in terraces that follow the natural contours of the escarpments like contour ploughing is a classic feature of the island of Bali and the Banaue Rice Terraces in Banaue, Ifugao, Philippines. In Peru, the Inca made use of otherwise unusable slopes by drystone walling to create terraces.

Terrace rice fields in Yunnan Province, China

Rice Paddies

A paddy field is a flooded parcel of arable land used for growing rice and other semiaquatic crops. Paddy fields are a typical feature of rice-growing countries of east and southeast Asia including Malaysia, China, Sri Lanka, Myanmar, Thailand, Korea, Japan, Vietnam, Taiwan, Indonesia, India, and the Philippines. They are also found in other rice-growing regions such as Piedmont (Italy), the Camargue (France) and the Artibonite Valley (Haiti). They can occur naturally along rivers or marshes, or can be constructed, even on hillsides. They require large water quantities for irrigation, much of it from flooding. It gives an environment favourable to the strain of rice being grown, and is hostile to many species of weeds. As the only draft animal species which is comfortable in wetlands, the water buffalo is in widespread use in Asian rice paddies.

Paddy-based rice-farming has been practiced in Korea since ancient times. A pit-house at the Daecheon-ni archaeological site yielded carbonized rice grains and radiocarbon dates indicating that rice cultivation may have begun as early as the Middle Jeulmun Pottery Period (c. 3500-2000 BC) in the Korean Peninsula. The earliest rice cultivation there may have used dry-fields instead of paddies.

The earliest Mumun features were usually located in naturally swampy, low-lying narrow gulleys and fed by local streams. Some Mumun paddies in flat areas were made of a series of squares and rectangles separated by bunds approximately 10 cm in height, while terraced paddies consisted of long irregularly shapes that followed natural contours of the land at various levels.

Like today's, Mumun period rice farmers used terracing, bunds, canals and small reservoirs. Some paddy-farming techniques of the Middle Mumun (c. 850-550 BC) can be interpreted from the well-preserved wooden tools excavated from archaeological rice paddies at the Majeon-ni Site. However, iron tools for paddy-farming were not introduced until sometime after 200 BC. The spatial scale of individual paddies, and thus entire paddy-fields, increased with the regular use of iron tools in the Three Kingdoms of Korea Period (c. AD 300/400-668).

A recent development in the intensive production of rice is System of Rice Intensification (SRI). Developed in 1983 by the French Jesuit Father Henri de Laulanié in Madagascar, by 2013 the number of smallholder farmers using SRI had grown to between 4 and 5 million.

Aquaculture

Aquaculture is the cultivation of the natural products of water (fish, shellfish, algae, seaweed and other aquatic organisms). Intensive aquaculture takes place on land using tanks, ponds or other controlled systems or in the ocean, using cages.

Sustainable Intensive Farming

Sustainable intensive farming practises have been developed to slow the deterioration of agricultural land and even regenerate soil health and ecosystem services, while still offering high yields. Most of these developments fall in the category of organic farming, or the integration of organic and conventional agriculture.

"Organic systems and the practices that make them effective are being picked up more and more by conventional agriculture and will become the foundation for future farming systems. They won't be called organic, because they'll still use some chemicals and still use some fertilizers, but they'll function much more like today's organic systems than today's conventional systems."

Dr. Charles Benbrook Executive director US House Agriculture Subcommittee Director Agricultural Board - National Academy Sciences (FMR)

The System of Crop Intensification (SCI) was born out of research primarily at Cornell University and smallholder farms in India on SRI. It uses the SRI concepts and methods for rice and applies them to crops like wheat, sugarcane, finger millet, and others. It can be 100% organic, or integrated with reduced conventional inputs.

Holistic management is a systems thinking approach that was originally developed for reversing desertification. Holistic planned grazing is similar to rotational grazing but differs in that it more explicitly provides a framework for adapting to four basic ecosystem processes: the water cycle, the mineral cycle including the carbon cycle, energy flow and community dynamics (the relationship between organisms in an ecosystem) as equal in importance to livestock production and social welfare. By intensively managing the behavior and movement of livestock, holistic planned grazing simultaneously increases stocking rates and restores grazing land.

Pasture cropping plants grain crops directly into grassland without first applying herbicides. The perennial grasses form a living mulch understory to the grain crop, eliminating the need to plant cover crops after harvest. The pasture is intensively grazed both before and after grain production using holistic planned grazing. This intensive system yields equivalent farmer profits (partly from increased livestock forage) while building new topsoil and sequestering up to 33 tons of CO_2/ha/year.

The Twelve Aprils grazing program for dairy production, developed in partnership with USDA-SARE, is similar to pasture cropping, but the crops planted into the perennial pasture are forage crops for dairy herds. This system improves milk production and is more sustainable than confinement dairy production.

Integrated Multi-Trophic Aquaculture (IMTA) is an example of a holistic approach. IMTA is a practice in which the by-products (wastes) from one species are recycled to become inputs (fertilizers, food) for another. Fed aquaculture (e.g. fish, shrimp) is combined with inorganic extractive (e.g. seaweed) and organic extractive (e.g. shellfish) aquaculture to create balanced systems for environmental sustainability (biomitigation), economic stability (product diversification and risk reduction) and social acceptability (better management practices).

Biointensive agriculture focuses on maximizing efficiency such as per unit area, energy input and water input. Agroforestry combines agriculture and orchard/forestry technologies to create more integrated, diverse, productive, profitable, healthy and sustainable land-use systems.

Intercropping can increase yields or reduce inputs and thus represents (potentially sustainable) agricultural intensification. However, while total yield per acre is often increased dramatically, yields of any single crop often diminish. There are also challenges to farmers relying on farming equipment optimized for monoculture, often resulting in increased labor inputs.

Vertical farming is intensive crop production on a large scale in urban centers in multi-story, artificially-lit structures that uses far less inputs and produces fewer environmental impacts.

An integrated farming system is a progressive biologically integrated sustainable agriculture system such as IMTA or Zero waste agriculture whose implementation requires exacting knowledge of the interactions of multiple species and whose benefits include sustainability and increased profitability. Elements of this integration can include:

- Intentionally introducing flowering plants into agricultural ecosystems to increase pollen-and nectar-resources required by natural enemies of insect pests

- Using crop rotation and cover crops to suppress nematodes in potatoes

Challenges

The challenges and issues of industrial agriculture for society, for the industrial agriculture sector, for the individual farm, and for animal rights include the costs and benefits of both current practices and proposed changes to those practices. This is a continuation of thousands of years of invention in feeding ever growing populations.

[W]hen hunter-gatherers with growing populations depleted the stocks of game and wild foods across the Near East, they were forced to introduce agriculture. But agriculture brought much longer hours of work and a less rich diet than hunter-gatherers enjoyed. Further population growth among shifting slash-and-burn farmers led to shorter fallow periods, falling yields and soil erosion. Plowing and fertilizers were introduced to deal with these problems - but once again involved longer hours of work and degradation of soil resources(Boserup, The Conditions of Agricultural Growth, Allen and Unwin, 1965, expanded and updated in Population and Technology, Blackwell, 1980.).

While the point of industrial agriculture is to profitably supply the world at the lowest cost, industrial methods have significant side effects. Further, industrial agriculture is not an indivisible whole, but instead is composed of multiple elements, each of which can be modified in response to market conditions, government regulation and further innovation and has its own side-effects. Various interest groups reach different conclusions on the subject.

Benefits

Population Growth

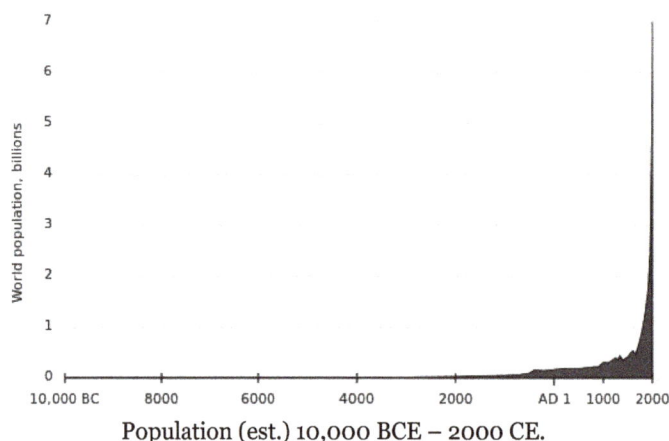

Population (est.) 10,000 BCE – 2000 CE.

Very roughly:

- 30,000 years ago hunter-gatherer behavior fed 6 million people

- 3,000 years ago primitive agriculture fed 60 million people

- 300 years ago intensive agriculture fed 600 million people

- Today industrial agriculture attempts to feed 6 billion people

Estimated world population at various dates, in **thousands**								
Year	**World**	**Africa**	**Asia**	**Europe**	**Central & South America**	**North America***	**Oceania**	**Notes**
8000 BCE	8 000							
1000 BCE	50 000							
500 BCE	100 000							
1 CE	200,000 plus							
1000	310 000							
1750	791 000	106 000	502 000	163 000	16 000	2 000	2 000	
1800	978 000	107 000	635 000	203 000	24 000	7 000	2 000	
1850	1 262 000	111 000	809 000	276 000	38 000	26 000	2 000	
1900	1 650 000	133 000	947 000	408 000	74 000	82 000	6 000	
1950	2 518 629	221 214	1 398 488	547 403	167 097	171 616	12 812	
1955	2 755 823	246 746	1 541 947	575 184	190 797	186 884	14 265	
1960	2 981 659	277 398	1 674 336	601 401	209 303	204 152	15 888	
1965	3 334 874	313 744	1 899 424	634 026	250 452	219 570	17 657	
1970	3 692 492	357 283	2 143 118	655 855	284 856	231 937	19 443	
1975	4 068 109	408 160	2 397 512	675 542	321 906	243 425	21 564	
1980	4 434 682	469 618	2 632 335	692 431	361 401	256 068	22 828	
1985	4 830 979	541 814	2 887 552	706 009	401 469	269 456	24 678	
1990	5 263 593	622 443	3 167 807	721 582	441 525	283 549	26 687	
1995	5 674 380	707 462	3 430 052	727 405	481 099	299 438	28 924	
2000	6 070 581	795 671	3 679 737	727 986	520 229	315 915	31 043	
2005	6 453 628	887 964	3 917 508	724 722	558 281	332 156	32 998**	

An example of industrial agriculture providing cheap and plentiful food is the U.S.'s "most successful program of agricultural development of any country in the world". Between 1930 and 2000 U.S. agricultural productivity (output divided by all inputs) rose by an average of about 2 percent annually causing food prices to decrease. "The percentage of U.S. disposable income spent on food prepared at home decreased, from 22 percent as late as 1950 to 7 percent by the end of the century."

Liabilities

Environment

Industrial agriculture uses huge amounts of water, energy, and industrial chemicals; increasing pollution in the arable land, usable water and atmosphere. Herbicides, insecticides and fertilizers are accumulating in ground and surface waters. "Many of the negative effects of industrial agriculture are remote from fields and farms. Nitrogen compounds from the Midwest, for example, travel down the Mississippi to degrade coastal fisheries in the Gulf of Mexico. But other adverse effects are showing up within agricultural production systems -- for example, the rapidly developing resistance among pests is rendering our arsenal of herbicides and insecticides increasingly ineffective.". Agrochemicals and monoculture have been implicated in Colony Collapse Disorder, in which the individual members of bee colonies disappear. Agricultural production is highly dependent on bees to pollinate many varieties of fruits and vegetables.

Social

A study done for the US. Office of Technology Assessment conducted by the UC Davis Macrosocial Accounting Project concluded that industrial agriculture is associated with substantial deterioration of human living conditions in nearby rural communities.

Modern Form of Intensive Farming

Intensive crop farming

Intensive crop farming is a modern form of intensive farming that refers to the industrialized production of crops. Intensive crop farming's methods include innovation in agricultural machinery, farming methods, genetic engineering technology, techniques for achieving economies of scale in production, the creation of new markets for consumption, patent protection of genetic information, and global trade. These methods are widespread in developed nations.

The practice of industrial agriculture is a relatively recent development in the history of agriculture, and the result of scientific discoveries and technological advances. Innovations in agriculture beginning in the late 19th century generally parallel developments in mass production in other industries that characterized the latter part of the Industrial Revolution. The identification of nitrogen and phosphorus as critical factors in plant growth led to the manufacture of synthetic fertilizers, making more intensive uses of farmland for crop production possible.

Similarly, the discovery of vitamins and their role in animal nutrition, in the first two decades of the 20th century, led to vitamin supplements, which in the 1920s allowed certain livestock to be raised indoors, reducing their exposure to adverse natural elements. The discovery of antibiotics and vaccines facilitated raising livestock in larger numbers by reducing disease. Chemicals developed for use in World War II gave rise to synthetic pesticides. Developments in shipping networks and technology have made long-distance distribution of produce feasible.

Crops

Features

- Large scale – hundreds or thousands of acres of a single crop (much more than can be absorbed into the local or regional market);

- Monoculture – large areas of a single crop, often raised from year to year on the same land, or with little crop rotation;

- Agrichemicals – reliance on imported, synthetic fertilizers and pesticides to provide nutrients and to mitigate pests and diseases, these applied on a regular schedule

- Hybrid seed – use of specialized hybrids designed to favor large scale distribution (e.g. ability to ripen off the vine, to withstand shipping and handling);

- Genetically engineered crops – use of genetically modified varieties designed for large scale production (e.g. ability to withstand selected herbicides);

- Large scale irrigation – heavy water use, and in some cases, growing of crops in otherwise unsuitable regions by extreme use of water (e.g. rice paddies on arid land).

- High mechanization – automated machinery sustain and harvest crops.

Criticism

Critics of intensively farmed crops cite a wide range of concerns. On the food quality front, it is held by critics that quality is reduced when crops are bred and grown primarily for cosmetic and shipping characteristics. Environmentally, industrial farming of crops is claimed to be responsible for loss of biodiversity, degradation of soil quality, soil erosion, food toxicity (pesticide residues) and pollution (through agrichemical build-ups and runoff, and use of fossil fuels for agrichemical manufacture and for farm machinery and long-distance distribution).

History

The projects within the Green Revolution spread technologies that had already existed, but had not been widely used outside of industrialized nations. These technologies included pesticides, irrigation projects, and synthetic nitrogen fertilizer.

The novel technological development of the Green Revolution was the production of what some referred to as "miracle seeds." Scientists created strains of maize, wheat, and rice that are generally referred to as HYVs or "high-yielding varieties." HYVs have an increased nitrogen-absorbing potential compared to other varieties. Since cereals that absorbed extra nitrogen would typically lodge, or fall over before harvest, semi-dwarfing genes were bred into their genomes. Norin 10 wheat, a variety developed by Orville Vogel from Japanese dwarf wheat varieties, was instrumental in developing Green Revolution wheat cultivars. IR8, the first widely implemented HYV rice to be developed by IRRI, was created through a cross between an Indonesian variety named "Peta" and a Chinese variety named "Dee Geo Woo Gen."

With the availability of molecular genetics in Arabidopsis and rice the mutant genes responsible (*reduced height(rht), gibberellin insensitive (gai1)* and *slender rice (slr1)*) have been cloned and identified as cellular signalling components of gibberellic acid, a phytohormone involved in regulating stem growth via its effect on cell division. Stem growth in the mutant background is significantly reduced leading to the dwarf phenotype. Photosynthetic investment in the stem is reduced dramatically as the shorter plants are inherently more stable mechanically. Assimilates become redirected to grain production, amplifying in particular the effect of chemical fertilisers on commercial yield.

HYVs significantly outperform traditional varieties in the presence of adequate irrigation, pesticides, and fertilizers. In the absence of these inputs, traditional varieties may outperform HYVs. One criticism of HYVs is that they were developed as F1 hybrids, meaning they need to be purchased by a farmer every season rather than saved from previous seasons, thus increasing a farmer's cost of production.

Examples

Wheat (Modern Management Techniques)

Wheat is a grass that is cultivated worldwide. Globally, it is the most important human food grain and ranks second in total production as a cereal crop behind maize; the third being rice. Wheat and barley were the first cereals known to have been domesticated. Cultivation and repeated harvesting and sowing of the grains of wild grasses led to the domestication of wheat through selection of mutant forms with tough ears which remained intact during harvesting, and larger grains. Because of the loss of seed dispersal mechanisms, domesticated wheats have limited capacity to propagate in the wild.

Agricultural cultivation using horse collar leveraged plows (3000 years ago) increased cereal grain productivity yields, as did the use of seed drills which replaced broadcasting sowing of seed in the 18th century. Yields of wheat continued to increase, as new land came under cultivation and with improved agricultural husbandry involving the use of fertilizers, threshing machines and reaping machines (the 'combine harvester'), tractor-draw cultivators and planters, and better varieties. With population growth rates falling, while yields continue to rise, the area devoted to wheat may now begin to decline for the first time in modern human history.

Organic wheat typically halves yield attainable but costs less as there are no fertiliser and pesticide costs. Seed costs are typically higher, however, and arguably labour and machinery costs are higher as the organic crop, and more importantly the whole rotation and cropping on such a farm, is more difficult to manage correctly.

While winter wheat lies dormant during a winter freeze, wheat normally requires between 110 and 130 days between planting and harvest, depending upon climate, seed type, and soil conditions. Crop management decisions require the knowledge of stage of development of the crop. In particular, spring fertilizers applications, herbicides, fungicides, growth regulators are typically applied at specific stages of plant development. For example, current recommendations often indicate the second application of nitrogen be done when the ear (not visible at this stage) is about 1 cm in size (Z31 on Zadoks scale).

Maize (mechanical harvesting)

Maize was planted by the Native Americans in hills, in a complex system known to some as the Three Sisters: beans used the corn plant for support, and squashes provided ground cover to stop weeds. This method was replaced by single species hill planting where each hill 60–120 cm (2–4 ft) apart was planted with 3 or 4 seeds, a method still used by home gardeners. A later technique was *checked corn* where hills were placed 40 inches (1,000 mm) apart in each direction, allowing cultivators to run through the field in two directions. In more arid lands this was altered and seeds were planted in the bottom of 10–12 cm (4–5 in) deep furrows to collect water. Modern technique plants maize in rows which allows for cultivation while the plant is young, although the hill technique is still used in the cornfields of some Native American reservations. Haudenosaunee Confederacy is what a group of Native Americans who are preparing for climate change through seed banking. Now this group is known as the Iroquois.

With a climate changing more crops are able to grow in different areas that they previously weren't able to grow in. This will open growing areas for maize.

A corn heap at the harvest site, India

In North America, fields are often planted in a two-crop rotation with a nitrogen-fixing crop, often alfalfa in cooler climates and soybeans in regions with longer summers. Sometimes a third crop, winter wheat, is added to the rotation. Fields are usually plowed each year, although no-till farming is increasing in use. Many of the maize varieties grown in the United States and Canada are hybrids. Over half of the corn area planted in the United States has been genetically modified using biotechnology to express agronomic traits such as pest resistance or herbicide resistance.

Before about World War II, most maize in North America was harvested by hand (as it still is in most of the other countries where it is grown). This often involved large numbers of workers and associated social events. Some one- and two-row mechanical pickers were in use but the corn combine was not adopted until after the War. By hand or mechanical picker, the entire ear is harvested which then requires a separate operation of a corn sheller to remove the kernels from the ear. Whole ears of corn were often stored in *corn cribs* and these whole ears are a sufficient form for some livestock feeding use. Few modern farms store maize in this manner. Most harvest the grain from the field and store it in bins. The combine with a corn head (with points and snap rolls instead of a reel) does not cut the stalk; it simply pulls the stalk down. The stalk continues downward and is crumpled into a mangled pile on the ground. The ear of corn is too large to pass through a slit in a plate and the snap rolls pull the ear of corn from the stalk so that only the ear and husk enter the machinery. The combine separates the husk and the cob, keeping only the kernels.

Soybean (Genetic Modification)

Soybeans are one of the "biotech food" crops that are being genetically modified, and GMO soybeans are being used in an increasing number of products. Monsanto Company is the world's leader in genetically modified soy for the commercial market. In 1995, Monsanto introduced "Roundup Ready" (RR) soybeans that have had a copy of a gene from the bacterium, *Agrobacterium* sp. strain CP4, inserted, by means of a gene gun, into its genome that allows the transgenic plant to survive being sprayed by this non-selective herbicide, glyphosate. Glyphosate, the active ingredient in Roundup, kills conventional soybeans. The bacterial gene is EPSP (= 5-enolpyruvyl shikimic acid-3-phosphate) synthase. Soybean also has a version of this gene, but the soybean version is sensitive to glyphosate, while the CP4 version is not.

RR soybeans allow a farmer to reduce tillage or even to sow the seed directly into an unplowed field, known as 'no-till' or conservation tillage. No-till agriculture has many advantages, greatly reducing soil erosion and creating better wildlife habitat; it also saves fossil fuels, and sequesters CO2, a greenhouse effect gas.

In *1997*, about 8% of all soybeans cultivated for the commercial market in the United States were genetically modified. In 2006, the figure was 89%. As with other "Roundup Ready" crops, concern is expressed over damage to biodiversity. However, the RR gene has been bred into so many different soybean cultivars that the genetic modification itself has not resulted in any decline of genetic diversity.

Tomato (Hydroponics)

The largest commercial hydroponics facility in the world is Eurofresh Farms in Willcox, Arizona, which sold more than 200 million pounds of tomatoes in 2007. Eurofresh has 318 acres (1.3 km²) under glass and represents about a third of the commercial hydroponic greenhouse area in the U.S. Eurofresh does not consider their tomatoes organic, but they are pesticide-free. They are grown in rockwool with top irrigation.

Some commercial installations use no pesticides or herbicides, preferring integrated pest management techniques. There is often a price premium willingly paid by consumers for produce which is labeled "organic". Some states in the USA require soil as an essential to obtain organic certification. There are also overlapping and somewhat contradictory rules established by the US Federal Government. So some food grown with hydroponics can be certified organic. In fact, they are the cleanest plants possible because there is no environment variable and the dirt in the food supply is extremely limited. Hydroponics also saves an incredible amount of water; It uses as little as 1/20 the amount as a regular farm to produce the same amount of food. The water table can be impacted by the water use and run-off of chemicals from farms, but hydroponics may minimize impact as well as having the advantage that water use and water returns are easier to measure. This can save the farmer money by allowing reduced water use and the ability to measure consequences to the land around a farm.

The environment in a hydroponics greenhouse is tightly controlled for maximum efficiency and this new mindset is called soil-less/controlled-environment agriculture (S/CEA). With this growers can make ultra-premium foods anywhere in the world, regardless of temperature and growing seasons. Growers monitor the temperature, humidity, and pH level constantly.

Vertical Farming

Vertical farming is the practice of producing food in vertically stacked layers, vertically inclined surfaces and/or integrated in other structures. The modern idea of vertical farming uses controlled-environment agriculture (CEA) technology, where all environmental factors can be controlled. These facilities utilize artificial control of light, environmental control (humidity, temperature, gases...) and fertigation. Some vertical farms use techniques similar to greenhouses, where natural sunlight can be augmented with artificial lighting and metal reflectors.

Lettuce grown in indoor vertical farming system.

Types

The term "vertical farming" was coined by Gilbert Ellis Bailey in 1915 in his book *Vertical Farming*. His use of the term differs from the current meaning—he wrote about farming with a special interest in soil origin, its nutrient content and the view of plant life as "vertical" life forms, specifically relating to root structures underground. Modern usage refers to skyscrapers using some degree of natural light.

Mixed-Use Skyscrapers

Mixed-use skyscrapers were proposed and built by architect Ken Yeang. Yeang proposes that instead of hermetically sealed mass-produced agriculture that plant life should be cultivated within open air, mixed-use skyscrapers for climate control and consumption (i.e., a personal or communal planting space as per the needs of the individual). This version of vertical farming is based upon personal or community use rather than the wholesale production and distribution plant life that aspires to feed an entire city. It thus requires less of an initial investment than Despommier's "vertical farm". However, neither Despommier nor Yeang are the conceptual originators, nor is Yeang the inventor of vertical farming in skyscrapers.

Plantagon is building the world's first industrial-scale multifunctional vertical farm in Linköping, Sweden.

Despommier's skyscrapers

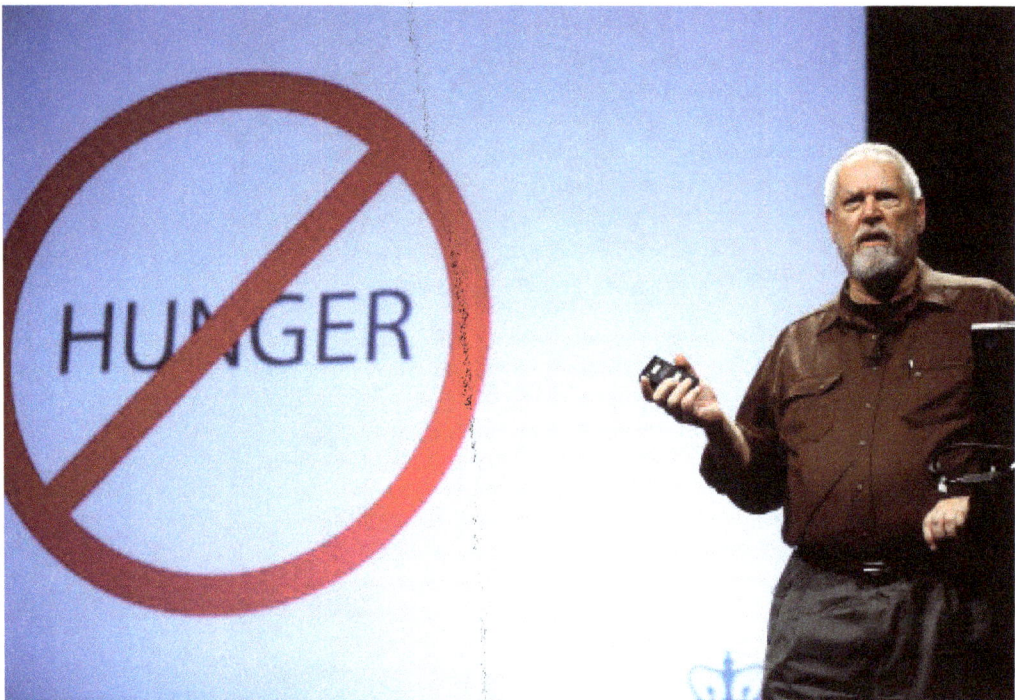

Dickson Despommier shares his ideas about how "vertical farming" can help reduce hunger by changing the way we use land for agriculture. Photography by Kris Krüg

Ecologist Dickson Despommier argues that vertical farming is legitimate for environmental reasons. He claims that the cultivation of plant life within skyscrapers will require less embedded energy and produce less pollution than some methods of producing plant life on natural landscapes. He moreover claims that natural landscapes are too toxic for natural, agricultural production, despite the ecological and environmental costs of extracting materials to build skyscrapers for the simple purpose of agricultural production.

Vertical farming according to Despommier thus discounts the value of natural landscape in exchange for the idea of "skyscraper as spaceship." Plant life is mass-produced within hermetically sealed, artificial environments that have little to do with the outside world. In this sense, they could be built anywhere regardless of the context. Although climate control, lighting, and other costs of maintenance have been posited as potentially stifling to bringing this concept to fruition, advocates have countered that an important feature of future vertical farms will be the integration of renewable energy technology, be it solar panels, wind turbines, water capture systems, and probably some combination of the three. The vertical farm is designed to be sustainable, and to enable nearby inhabitants to work at the farm.

Despommier's concept of the vertical farm emerged in 1999 at Columbia University. It promotes the mass cultivation of plant life for commercial purposes in skyscrapers.

Stackable Shipping Containers

Several companies have brought forth the concept of stacking recycled shipping containers in urban settings. Freight Farms produces a "leafty green machine" that is a complete farm-to-table system outfitted with vertical hydroponics, LED lighting and intuitive climate controls built within a 12 m × 2.4 m shipping container. Podponics has built a large scale vertical farm in Atlanta consisting of over 100 stacked "growpods". A similar farm is currently under construction in Oman.

History

A commercial high-rise farm such as the 'Vertical Farm' has never been built, yet extensive photographic documentation and several historical books on the subject suggest that research on the subject was not diligently pursued. New sources indicate that a tower hydroponicum existed in Armenia prior to 1951.

Proponents argue that, by allowing traditional outdoor farms to revert to a natural state and reducing the energy costs needed to transport foods to consumers, vertical farms could significantly alleviate climate change produced by excess atmospheric carbon. Critics have noted that the costs of the additional energy needed for artificial lighting, heating and other vertical farming operations would outweigh the benefit of the building's close proximity to the areas of consumption. However, a recent study published in the Journal of Agricultural Engineering and Biotechnology has utilized inexpensive metal reflectors to supply sunlight to the plants.

One of the earliest drawings of a tall building that cultivates food was published in Life Magazine in 1909. The reproduced drawings feature vertically stacked homesteads set amidst a farming landscape. This proposal can be seen in Rem Koolhaas's Delirious New York. Koolhaas wrote that this 1909 theorem is 'The Skyscraper as Utopian device for the production of unlimited numbers of virgin sites on a metropolitan location' (1994, 82).

Other architectural proposals that provide the seeds for the Vertical Farm project include Le Corbusier's Immeubles-Villas (1922) and SITE's Highrise of homes (1972). SITE's Highrise of homes, is a near revival of the 1909 Life Magazine Theorem. In fact, built examples of tower hydroponicums are quite well documented in the canonical text of "The Glass House" by John Hix. Images of the vertical farms at the School of Gardeners in Langenlois, Austria, and the glass tower at the Vienna International Horticulture Exhibition (1964) clearly show that vertical farms existed more than 40 years prior to contemporary discourse on the subject. Although architectural precedents remain valuable, the technological precedents that make vertical farming possible can be traced back to horticultural history through the development of greenhouse and hydroponic technology. Early building types or Hydroponicums were developed, integrating hydroponic technology into building systems. These horticultural building systems evolved from greenhouse technology, and paved the way for the modern concept of the vertical farm. The British Interplanetary Society developed a hydroponicum for lunar conditions and other building prototypes were developed during the early days of space exploration. During this era of expansion and experimentation, the first Tower Hydroponic Units were developed in Armenia.

The Armenian tower hydroponicums are the first built examples of a vertical farm, and is documented in Sholto Douglas' seminal text "Hydroponics: The Bengal System" first published in 1951 with data from the then-East Pakistan, today's Bangladesh, and the Indian state of West Bengal. Contemporary notions of vertical farming are predated by this early technology by more than 50 years. Contemporary precursors that have been published, or built, are Ken Yeang's Bioclimatic Skyscraper (Menara Mesiniaga, built 1992); MVRDV's PigCity, 2000; MVRDV's Meta City/ Datatown (1998–2000); Pich-Aguilera's Garden Towers (2001).

Ken Yeang is perhaps the most widely known architects that has promoted the idea of the 'mixed-use' Bioclimatic Skyscraper which combines living units and opportunities for food production.

Early prototypes of vertical farms, or "Tower Hydroponicums" existed in Armenia prior to 1951 during an era of hydroponic and horticultural building system research fueled by space exploration and a transatlantic technology race.

The latest version of these very idea is Dickson Despommier's "The Vertical Farm".

Dickson Despommier, a professor of environmental health sciences and microbiology at Columbia University in New York City, modernized the idea of vertical farming in 1999 with graduate students in a medical ecology class. Although much of Despommier's suggestions have been challenged and strongly criticized from an environmental science and engineering point of view, the idea's popularization in recent years has been largely the result of Despommier's assertion that food production can be transformed.

Despommier had originally challenged his class to feed the entire population of Manhattan (About 2,000,000 people) using only 5 hectares (13 acres) of usable rooftop gardens. The class calculated that, by using rooftop gardening methods, only 2 percent of the population would be fed. Unsatisfied with the results, Despommier made an off-the-cuff suggestion of growing plants indoors, vertically. The idea sparked the students' interests and gained major momentum. By 2001 the first outline of a vertical farm was introduced and today scientists, architects, and investors worldwide are working together to make the concept of vertical farming a reality. In an interview with Miller-McCune.com, Despommier described how vertical farms would function:

> Each floor will have its own watering and nutrient monitoring systems. There will be sensors for every single plant that tracks how much and what kinds of nutrients the plant has absorbed. You'll even have systems to monitor plant diseases by employing DNA chip technologies that detect the presence of plant pathogens by simply sampling the air and using snippets from various viral and bacterial infections. It's very easy to do.
>
> Moreover, a gas chromatograph will tell us when to pick the plant by analyzing which flavenoids the produce contains. These flavonoids are what gives the food the flavors you're so fond of, particularly for more aromatic produce like tomatoes and peppers. These are all right-off-the-shelf technologies. The ability to construct a vertical farm exists now. We don't have to make anything new.

Architectural designs have been produced by Chris Jacobs and Andrew Kranis from Columbia University and Gordon Graff from the University of Waterloo's School of Architecture in Cambridge, ON.

Mass media attention began with an article written in *New York* magazine. Since 2007, articles have appeared in *The New York Times*, *U.S. News & World Report*, *Popular Science*, *Scientific American* and *Maxim*, among others, as well as radio and television features.

As of 2012, Vertical Harvest is working on raising funds for an urban, small-scale vertical farm in Jackson Hole, Wyoming.

As of 2014, Vertical Fresh Farms has been operating in Buffalo, NY specializing in a wide variety of salad greens, herbs, and sprouts.

Problems

Economics

Opponents question the potential profitability of vertical farming. A detailed cost analysis of start-up costs, operation costs and revenue has not been done. The extra cost of lighting, heating, and powering the vertical farm may negate any of the cost benefits received by the decrease in transportation expenses. The economic and environmental benefits of vertical farming rest partly on the concept of minimizing food miles, the distance that food travels from farm to consumer. However, a recent analysis suggests that transportation is only a minor contributor to the economic and environmental costs of supplying food to urban populations. The author of the report, University of Toronto professor Pierre Desrochers, concluded that "food miles are, at best, a marketing fad." Thus the facility would have to produce a considerable profit to justify remaining in the city. A simpler concept rather than trying to stack farms on multiple stories would be to just cultivate crops on the roofs of existing building. Rooftop farming is a growing urban trend, requires little construction (other than fortifying the roof to hold the weight of the growing medium), still takes advantage of sunlight and doesn't require investment in machinery, growing lights or irrigation.

Similarly, if the power needs of the vertical farm are met by fossil fuels, the environmental effect may be a net loss; even building low-carbon capacity to power the farms may not make as much sense as simply leaving the traditional farms in place, and burning less coal.

The initial building costs will be easily over $100 million, for a 60 hectare vertical farm. Office occupancy costs can be very high in major cities, with cities such as Tokyo, Moscow, Mumbai, Dubai, Milan, Zurich, and Sao Paulo ranging from $1850 to $880 per square meter, respectively.

Energy use

During the growing season, the sun shines on a vertical surface at an extreme angle such that much less light is available to crops than when they are planted on flat land. Therefore, supplemental light would be required in order to obtain economically viable yields. Bruce Bugbee, a crop physiologist at Utah State University, believes that the power demands of vertical farming will be too expensive and uncompetitive with traditional farms using only free natural light. The environmental writer George Monbiot calculated that the cost of providing enough supplementary light to grow the grain for a single loaf would be about $15. An article in the Economist argued that "even though crops growing in a glass skyscraper will get some natural sunlight during the day, it won't be enough" and "the cost of powering artificial lights will make indoor farming prohibitively expensive".

As "The Vertical Farm" proposes a controlled environment, heating and cooling costs will be at least as costly as any other tower. But there also remains the issue of complicated, if not more expensive, plumbing and elevator systems to distribute food and water throughout. Even throughout the northern continental United States, while heating with relatively cheap fossil fuels, the heating cost can be over $200,000 per hectare.

To address this problem, The Plant in Chicago is building an anaerobic digester into the building. This will allow the farm to operate off the energy grid. Moreover, the anaerobic digester will be recycling waste from nearby businesses that would otherwise go into landfills.

Pollution

Depending on the method of electricity generation used, regular greenhouse produce can create more greenhouse gases than field produce, largely due to higher energy use per kilogram of produce. With vertical farms requiring much greater energy per kilogram of produce, mainly through increased lighting, than regular greenhouses, the amount of pollution created will be much higher than that from field produce. The amount of pollution produced is dependent on how the energy used in the process is generated.

As plants acquire nearly all their carbon from the atmosphere, greenhouse growers commonly supplement CO_2 levels to 3–4 times the rate normally found in the atmosphere. This increase in CO_2, which has been shown to increase photosynthesis rates by 50%, contributes to the higher yields expected in vertical farming. It is not uncommon to find greenhouses burning fossil fuels purely for this purpose, as other CO_2 sources, like from furnaces, contain pollutants such as sulphur dioxide and ethylene which significantly damage plants. This means a vertical farm will require a CO_2 source, most likely from combustion, even if the rest of the farm is powered by "green" energy. Also, through necessary ventilation, much CO_2 will be leaked into the city's atmosphere.

Greenhouse growers commonly exploit photoperiodism in plants to control whether the plants are in a vegetative or reproductive stage. As part of this control, growers will have the lights on past sunset and before sunrise or periodically throughout the night. Single story greenhouses are already a nuisance to neighbours because of light pollution, a 30-story vertical farm in a densely populated area will surely face problems because of its light pollution.

Hydroponics greenhouses regularly change the water, meaning there is a large quantity of water containing fertilizers and pesticides that must be disposed of. While solutions are currently being worked on, the most common method of simply spreading the mixture over a sufficient area of neighbouring farmland or wetlands would be more difficult for an urban vertical farm.

Advantages

Several potential advantages of vertical farming have been discussed by Despommier. Many of these benefits are obtained from scaling up hydroponic or aeroponic growing methods.

Preparation for the Future

It is estimated that by the year 2050, close to 80% of the world's population will live in urban areas and the total population of the world will increase by 3 billion people. A very large amount of land may be required depending on the change in yield per hectare. Scientists are concerned that this large amount of required farmland will not be available and that severe damage to the earth will be caused by the added farmland. According to Despommier, vertical farms, if designed properly, may eliminate the need to create additional farmland and help create a cleaner environment.

Increased Crop Production

Unlike traditional farming in non-tropical areas, indoor farming can produce crops year-round. All-season farming multiplies the productivity of the farmed surface by a factor of 4 to 6 depending on the crop. With some crops, such as strawberries, the factor may be as high as 30.

Furthermore, as the crops would be sold in the same infrastructures in which they are grown, they will not need to be transported between production and sale, resulting in less spoilage, infestation, and energy required than conventional farming encounters. Research has shown that 30% of harvested crops are wasted due to spoilage and infestation, though this number is much lower in developed nations.

Despommier suggests that, if dwarf versions of certain crops are used (e.g. dwarf wheat, which has been grown in space by NASA, is smaller in size but richer in nutrients), year-round crops, and "stacker" plant holders are accounted for, a 30-story building with a base of a building block (2 hectares (5 acres)) would yield a yearly crop analogous to that of 1,000 hectares (2,400 acres) of traditional farming.

Protection from Weather-Related Problems

Crops grown in traditional outdoor farming suffer from the often suboptimal, and sometimes extreme, nature of geological and meteorological events such as undesirable temperatures or rainfall amounts, monsoons, hailstorms, tornadoes, flooding, wildfires, and severe droughts. The protection of crops from weather is increasingly important as global climate change occurs. "Three recent floods (in 1993, 2007 and 2008) cost the United States billions of dollars in lost crops, with even more devastating losses in topsoil. Changes in rain patterns and temperature could diminish India's agricultural output by 30 percent by the end of the century."

Because vertical plant farming provides a controlled environment, the productivity of vertical farms would be mostly independent of weather and protected from extreme weather events. Although the controlled environment of vertical farming negates most of these factors, earthquakes and tornadoes still pose threats to the proposed infrastructure, although this again depends on the location of the vertical farms.

Conservation of Resources

Each unit of area in a vertical farm could allow up to 20 units of area of outdoor farmland to return to its natural state, and recover farmlands due to development from original flat farmlands.

Vertical farming would reduce the need for new farmland due to overpopulation, thus saving many natural resources, currently threatened by deforestation or pollution. Deforestation and desertification caused by agricultural encroachment on natural biomes would be avoided. Because vertical farming lets crops be grown closer to consumers, it would substantially reduce the amount of fossil fuels currently used to transport and refrigerate farm produce. Producing food indoors reduces or eliminates conventional plowing, planting, and harvesting by farm machinery, also powered by fossil fuels.

Halting Mass Extinction

Withdrawing human activity from large areas of the Earth's land surface may be necessary to slow and eventually halt the current anthropogenic mass extinction of land animals.

Traditional agriculture is highly disruptive to wild animal populations that live in and around farmland and some argue it becomes unethical when there is a viable alternative. One study showed that wood mouse populations dropped from 25 per hectare to 5 per hectare after harvest, estimating 10 animals killed per hectare each year with conventional farming. In comparison, vertical farming would cause very little harm to wildlife, and would allow disused farmland to return to its pre-agricultural state.

Impact on Human Health

Traditional farming is a hazardous occupation with particular risks that often take their toll on the health of human laborers. Such risks include: exposure to infectious diseases such as malaria and schistosomes, exposure to toxic chemicals commonly used as pesticides and fungicides, confrontations with dangerous wildlife such as venomous snakes, and the severe injuries that can occur when using large industrial farming equipment. Whereas the traditional farming environment inevitably contains these risks (particularly in the farming practice known as "slash and burn"), vertical farming – because the environment is strictly controlled and predictable – reduces some of these dangers. Currently, the American food system makes fast, unhealthy food cheap while fresh produce is less available and more expensive, encouraging poor eating habits. These poor eating habits lead to health problems such as obesity, heart disease, and diabetes. The increased availability and subsequent lower cost of fresh produce would encourage healthier eating.

Poverty/Destitution and Culture

Food security is one of the primary factors leading to absolute poverty. Being able to construct 'farm land' in secure areas as needed will help alleviate the pressures causing crises among neighbors fighting for resources (mainly water and space). It also allows continued growth of culturally significant food items without sacrificing sustainability or basic needs, which can be significant to the recovery of a society from poverty.

Urban Growth

Vertical farming, used in conjunction with other technologies and socioeconomic practices, could allow cities to expand while remaining largely self-sufficient food wise. This would allow for large urban centers that could grow without destroying considerably larger areas of forest to provide food for their people. Moreover, the industry of vertical farming will provide employment to these expanding urban centers. This may help displace the unemployment created by the dismantling of traditional farms, as more farm laborers move to cities in search of work.

Energy Sustainability

Vertical farms could exploit methane digesters to generate a small portion of its own electrical needs. Methane digesters could be built on site to transform the organic waste generated at the farm into biogas which is generally composed of 65% methane along with other gases. This biogas could then be burned to generate electricity for the greenhouse.

Technologies and Devices

Vertical farming relies on the use of various physical methods to become effective. Combining these technologies and devices in an integrated whole is necessary to make Vertical Farming a reality. Various methods are proposed and under research. The most common technologies suggested are:

- Greenhouse
- The Folkewall and other vertical growing architectures
- Flowerpot
- Aeroponics / Hydroponics / Aquaponics
- Composting
- Grow light
- Phytoremediation

- Skyscraper

- Controlled-environment agriculture

- Precision agriculture

- Agricultural robot

Plans

Despommier argues that the technology to construct vertical farms currently exists. He also states that the system can be profitable and effective, a claim evidenced by some preliminary research posted on the project's website. Developers and local governments in the following cities have expressed serious interest in establishing a vertical farm: Incheon (South Korea), Abu Dhabi (United Arab Emirates), and Dongtan (China), New York City, Portland, Ore., Los Angeles, Las Vegas, Seattle, Surrey, B.C., Toronto, Paris, Bangalore, Dubai, Shanghai and Beijing. The Illinois Institute of Technology is now crafting a detailed plan for Chicago. It is suggested that prototype versions of vertical farms should be created first, possibly at large universities interested in the research of vertical farms, in order to prevent failures such as the Biosphere 2 project in Oracle, Arizona.

In 2009, the world's first pilot production system was installed at Paignton Zoo Environmental Park in the United Kingdom. The project showcased a technological solution for vertical farming and provided a physical base to conduct research into sustainable urban food production. The produce is used to feed the zoo's animals while the project enables evaluation of the systems and provides an educational resource to advocate for change in unsustainable land use practices that impact upon global biodiversity and ecosystem services,

In 2010, the Green Zionist Alliance proposed a resolution at the 36th World Zionist Congress calling on Keren Kayemet L'Yisrael (Jewish National Fund in Israel) to develop vertical farms in Israel.

In 2012, the world's first commercial vertical farm was opened in Singapore, developed by Sky Greens Farms, and is three stories high. They currently have over 100 towers that stand at nine meters tall.

In 2013 (July 18) the Association for Vertical Farming (AVF) was founded in Munich (Germany). By May 2015 the AVF already expanded with regional chapters all over Europe, Asia, USA, Canada and the United Kingdom. This internationally active non-profit organization unites growers and inventors to improve food security and the sustainable development of Vertical Farming. The AVF also focuses on advancing vertical farming technologies, designs and businesses by hosting international info-days, workshops and summits. They developed a glossary to bring consistency to the industry and plan on helping to standardize the technologies.

References

- "USDA List of Allowed and Prohibited Substances in Organic Agriculture". USDA List of Allowed and Prohibited Substances in Organic Agriculture. USDA. 4 April 2016. Retrieved 6 April 2016.

- Arsenault, Chris. "Only 60 Years of Farming Left If Soil Degradation Continues". Scientific American. Retrieved 29 May 2016.

- Paull, John "From France to the World: The International Federation of Organic Agriculture Movements (IFOAM)", Journal of Social Research & Policy, 2010, 1(2):93-102.

- Danielle Treadwell, Jim Riddle, Mary Barbercheck, Deborah Cavanaugh-Grant, Ed Zaborski, ''Cooperative Extension System", What is organic farming?

- Paull, John (2011) "The Uptake of Organic Agriculture: A Decade of Worldwide Development", Journal of Social and Development Sciences, 2 (3), pp. 111-120.

- Horne, Paul Anthony (2008). Integrated pest management for crops and pastures. CSIRO Publishing. p. 2. ISBN 978-0-643-09257-0.

- Paull, John (2011). "Attending the First Organic Agriculture Course: Rudolf Steiner's Agriculture Course at Koberwitz, 1924". European Journal of Social Sciences'. 21 (1): 64–70.

- Vogt G (2007). Lockeretz W, ed. Chapter 1: The Origins of Organic Farming. Organic Farming: An International History. CABI Publishing. pp. 9–30. ISBN 9780851998336.

Related Aspects of Sustainable Agriculture and Farming

The aim of this chapter is to provide the readers an in-depth understanding of sustainable agriculture by elucidating the fields that are closely related to it and play a role in the field's progress. Topics such as agroecology, agroforestry, conservation agriculture among others have been rigorously examined in this chapter.

Agroecology

A community-supported agriculture share of crops.

Agroecology is the study of ecological processes that operate in agricultural production systems. The prefix *agro-* refers to *agriculture*. Bringing ecological principles to bear in agroecosystems can suggest novel management approaches that would not otherwise be considered. The term is often used imprecisely and may refer to "a science, a movement, [or] a practice." Agroecologists study a variety of agroecosystems, and the field of agroecology is not associated with any one particular method of farming, whether it be organic, integrated, or conventional; intensive or extensive. Although it has much more common thinking and principles with some of the before mentioned farming systems.

Ecological Strategy

Agroecologists do not unanimously oppose technology or inputs in agriculture but instead assess how, when, and if technology can be used in conjunction with natural, social and human assets. Agroecology proposes a context- or site-specific manner of studying agroecosystems, and as such, it recognizes that there is no universal formula or recipe for the success and maximum well-being of an agroecosystem. Thus, agroecology is not defined by certain management practices, such as the use of natural enemies in place of insecticides, or polyculture in place of monoculture.

Instead, agroecologists may study questions related to the four system properties of agroecosystems: productivity, stability, sustainability and equitability. As opposed to disciplines that are concerned with only one or some of these properties, agroecologists see all four properties as interconnected and integral to the success of an agroecosystem. Recognizing that these properties are found on varying spatial scales, agroecologists do not limit themselves to the study of agroecosystems at any one scale: gene-organism-population-community-ecosystem-landscape-biome, field-farm-community-region-state-country-continent-global.

Agroecologists study these four properties through an interdisciplinary lens, using natural sciences to understand elements of agroecosystems such as soil properties and plant-insect interactions, as well as using social sciences to understand the effects of farming practices on rural communities, economic constraints to developing new production methods, or cultural factors determining farming practices.

Approaches

Agroecologists do not always agree about what agroecology is or should be in the long-term. Different definitions of the term agroecology can be distinguished largely by the specificity with which one defines the term "ecology," as well as the term's potential political connotations. Definitions of agroecology, therefore, may be first grouped according to the specific contexts within which they situate agriculture. Agroecology is defined by the OECD as "the study of the relation of agricultural crops and environment." This definition refers to the "-ecology" part of "agroecology" narrowly as the natural environment. Following this definition, an agroecologist would study agriculture's various relationships with soil health, water quality, air quality, meso- and micro-fauna, surrounding flora, environmental toxins, and other environmental contexts.

A more common definition of the word can be taken from Dalgaard et al., who refer to agroecology as the study of the interactions between plants, animals, humans and the environment within agricultural systems. Consequently, agroecology is inherently multidisciplinary, including factors from agronomy, ecology, sociology, economics and related disciplines. In this case, the "-ecology" portion of "agroecology is defined broadly to include social, cultural, and economic contexts as well. Francis et al. also expand the definition in the same way, but put more emphasis on the notion of food systems.

Agroecology is also defined differently according to geographic location. In the global south, the term often carries overtly political connotations. Such political definitions of the term usually ascribe to it the goals of social and economic justice; special attention, in this case, is often paid to the traditional farming knowledge of indigenous populations. North American and European uses of the term sometimes avoid the inclusion of such overtly political goals. In these cases, agroecology is seen more strictly as a scientific discipline with less specific social goals.

Agro-Population Ecology

This approach is derived from the science of ecology primarily based on population ecology, which over the past three decades has been displacing the ecosystems biology of Odum. Buttel explains the main difference between the two categories, saying that "the application of population ecology to agroecology involves the primacy not only of analyzing agroecosystems from the perspective of the population dynamics of their constituent species, and their relationships to climate and biogeochemistry, but also there is a major emphasis placed on the role of genetics."

Inclusive Agroecology

Rather than viewing agroecology as a subset of agriculture, Wojtkowski takes a more encompassing perspective. In this, natural ecology and agroecology are the major headings under ecology. Natural ecology is the study of organisms as they interact with and within natural environments. Correspondingly, agroecology is the basis for the land-use sciences. Here humans are the primary governing force for organisms within planned and managed, mostly terrestrial, environments.

As key headings, natural ecology and agroecology provide the theoretical base for their respective sciences. These theoretical bases overlap but differ in a major way. Economics has no role in the functioning of natural ecosystems whereas economics sets direction and purpose in agroecology.

Under agroecology are the three land-use sciences, agriculture, forestry, and agroforestry. Although these use their plant components in different ways, they share the same theoretical core.

Beyond this, the land-use sciences further subdivide. The subheadings include agronomy, organic farming, traditional agriculture, permaculture, and silviculture. Within this system of subdivisions, agroecology is philosophically neutral. The importance lies in providing a theoretical base hitherto lacking in the land-use sciences. This allows progress in biocomplex agroecosystems including the multi-species plantations of forestry and agroforestry.

Applications

To arrive at a point of view about a particular way of farming, an agroecologist would first seek to understand the contexts in which the farm(s) is(are) involved. Each farm may be inserted in a unique combination of factors or contexts. Each farmer may have their own premises about the meanings of an agricultural endeavor, and these meanings might be different from those of agroecologists. Generally, farmers seek a configuration that is viable in multiple contexts, such as family, financial, technical, political, logistical, market, environmental, spiritual. Agroecologists want to understand the behavior of those who seek livelihoods from plant and animal increase, acknowledging the organization and planning that is required to run a farm.

Views on Organic and Non-Organic Milk Production

Because organic agriculture proclaims to sustain the health of soils, ecosystems, and people, it has much in common with Agroecology; this does not mean that Agroecology is synonymous with organic agriculture, nor that Agroecology views organic farming as the 'right' way of farming. Also, it is important to point out that there are large differences in organic standards among countries and certifying agencies.

Three of the main areas that agroecologists would look at in farms, would be: the environmental impacts, animal welfare issues, and the social aspects.

Environmental impacts caused by organic and non-organic milk production can vary significantly. For both cases, there are positive and negative environmental consequences.

Compared to conventional milk production, organic milk production tends to have lower eutrophication potential per ton of milk or per hectare of farmland, because it potentially reduces leaching of nitrates (NO_3^-) and phosphates (PO_4^-) due to lower fertilizer application rates. Because organic milk production reduces pesticides utilization, it increases land use per ton of milk due to decreased crop yields per hectare. Mainly due to the lower level of concentrates given to cows in organic herds, organic dairy farms generally produce less milk per cow than conventional dairy farms. Because of the increased use of roughage and the, on-average, lower milk production level per cow, some research has connected organic milk production with increases in the emission of methane.

Animal welfare issues vary among dairy farms and are not necessarily related to the way of producing milk (organically or conventionally).

A key component of animal welfare is freedom to perform their innate (natural) behavior, and this is stated in one of the basic principles of organic agriculture. Also, there are other aspects of animal welfare to be considered - such as freedom from hunger, thirst, discomfort, injury, fear, distress, disease and pain. Because organic standards require loose housing systems, adequate bedding, restrictions on the area of slatted floors, a minimum forage proportion in the ruminant diets, and tend to limit stocking densities both on pasture and in housing for dairy cows, they potentially promote good foot and hoof health. Some studies show lower incidence of placenta retention, milk fever, abomasums displacement and other diseases in organic than in conventional dairy herds. However, the level of infections by parasites in organically managed herds is generally higher than in conventional herds.

Social aspects of dairy enterprises include life quality of farmers, of farm labor, of rural and urban communities, and also includes public health.

Both organic and non-organic farms can have good and bad implications for the life quality of all the different people involved in that food chain. Issues like labor conditions, labor hours and labor rights, for instance, do not depend on the organic/non-organic characteristic of the farm; they can be more related to the socio-economical and cultural situations in which the farm is inserted, instead.

As for the public health or food safety concern, organic foods are intended to be healthy, free of contaminations and free from agents that could cause human diseases. Organic milk is meant to have no chemical residues to consumers, and the restrictions on the use of antibiotics and chemicals in organic food production has the purpose to accomplish this goal. Although dairy cows in both organic and conventional farming practices can be exposed to pathogens, it has been shown that, because antibiotics are not permitted as a preventative measure in organic practices, there are far fewer antibiotic resistant pathogens on organic farms. This dramatically increases the efficacy of antibiotics when/if they are necessary.

In an organic dairy farm, an agroecologist could evaluate the following:

1. Can the farm minimize environmental impacts and increase its level of sustainability, for instance by efficiently increasing the productivity of the animals to minimize waste of feed and of land use?

2. Are there ways to improve the health status of the herd (in the case of organics, by using biological controls, for instance)?

3. Does this way of farming sustain good quality of life for the farmers, their families, rural labor and communities involved?

Views on No-Till Farming

No-tillage is one of the components of conservation agriculture practices and is considered more environmental friendly than complete tillage. There is a general consensus that no-till can increase soils capacity of acting as a carbon sink, especially when combined with cover crops.

No-till can contribute to higher soil organic matter and organic carbon content in soils, though reports of no-effects of no-tillage in organic matter and organic carbon soil contents also exist, depending on environmental and crop conditions. In addition, no-till can indirectly reduce CO_2 emissions by decreasing the use of fossil fuels.

Most crops can benefit from the practice of no-till, but not all crops are suitable for complete no-till agriculture. Crops that do not perform well when competing with other plants that grow in untilled soil in their early stages can be best grown by using other conservation tillage practices, like a combination of strip-till with no-till areas. Also, crops which harvestable portion grows underground can have better results with strip-tillage, mainly in soils which are hard for plant roots to penetrate into deeper layers to access water and nutrients.

The benefits provided by no-tillage to predators may lead to larger predator populations, which is a good way to control pests (biological control), but also can facilitate predation of the crop itself. In corn crops, for instance, predation by caterpillars can be higher in no-till than in conventional tillage fields.

In places with rigorous winter, untilled soil can take longer to warm and dry in spring, which may delay planting to less ideal dates. Another factor to be considered is that organic residue from the prior year's crops lying on the surface of untilled fields can provide a favorable environment to pathogens, helping to increase the risk of transmitting diseases to the future crop. And because no-till farming provides good environment for pathogens, insects and weeds, it can lead farmers to a more intensive use of chemicals for pest control. Other disadvantages of no-till include underground rot, low soil temperatures and high moisture.

Based on the balance of these factors, and because each farm has different problems, agroecologists will not atest that only no-till or complete tillage is the right way of farming. Yet, these are not the only possible choices regarding soil preparation, since there are intermediate practices such as strip-till, mulch-till and ridge-till, all of them - just as no-till - categorized as conservation tillage. Agroecologists, then, will evaluate the need of different practices for the contexts in which each farm is inserted.

In a no-till system, an agroecologist could ask the following:

1. Can the farm minimize environmental impacts and increase its level of sustainability; for instance by efficiently increasing the productivity of the crops to minimize land use?

2. Does this way of farming sustain good quality of life for the farmers, their families, rural labor and rural communities involved?

History

Pre-WWII

The notions and ideas relating to crop ecology have been around since at least 1911 when F.H. King released *Farmers of Forty Centuries*. King was one of the pioneers as a proponent of more quantitative methods for characterization of water relations and physical properties of soils. In the late 1920s the attempt to merge agronomy and ecology was born with the development of the field of crop ecology. Crop ecology's main concern was where crops would be best grown. Actually, it was only in 1928 that agronomy and ecology were formally linked by Klages.

The first mention of the term agroecology was in 1928, with the publication of the term by Bensin in 1928. The book of Tischler (1965), was probably the first to be actually titled 'agroecology'. He analysed the different components (plants, animals, soils and climate) and their interactions within an agroecosystem as well as the impact of human agricultural management on these components. Other books dealing with agroecology, but without using the term explicitly were published by the German zoologist Friederichs (1930) with his book on agricultural zoology and related ecological/environmental factors for plant protection, and by American crop physiologist Hansen in 1939 when both used the word as a synonym for the application of ecology within agriculture.

Post-WWII

Gliessman mentions that post-WWII, groups of scientists with ecologists gave more focus to experiments in the natural environment, while agronomists dedicated their attention to the cultivated systems in agriculture. According to Gliessman, the two groups kept their research and interest apart until books and articles using the concept of agroecosystems and the word agroecology started to appear in 1970. Dalgaard explains the different points of view in ecology schools, and the fundamental differences, which set the basis for the development of agroecology. The early ecology school of Henry Gleason investigated plant populations focusing in the hierarchical levels of the organism under study.

Friederich Clement's ecology school, however included the organism in question as well as the higher hierarchical levels in its investigations, a "landscape perspective". However, the ecological schools where the roots of agroecology lie are even broader in nature. The ecology school of Tansley, whose view included both the biotic organism and their environment, is the one from which the concept of agroecosystems emerged in 1974 with Harper.

In the 1960s and 1970s the increasing awareness of how humans manage the landscape and its consequences set the stage for the necessary cross between agronomy and ecology. Even though, in many ways the environmental movement in the US was a product of the times, the Green Decade, spread an environmental awareness of the unintended consequences of changing ecological processes. Works such as *Silent Spring*, and *The Limits to Growth*, and changes in legislation such as the Clean Air Act, Clean Water Act, and the National Environmental Policy Act caused the public to be aware of societal growth patterns, agricultural production, and the overall capacity of the system.

Fusion with Ecology

After the 1970s, when agronomists saw the value of ecology and ecologists began to use the agricultural systems as study plots, studies in agroecology grew more rapidly. Gliessman describes that the innovative work of Prof. Efraim Hernandez X., who developed research based on indigenous systems of knowledge in Mexico, led to education programs in agroecology. In 1977 Prof. Efraim Hernandez X. explained that modern agricultural systems had lost their ecological foundation when socio-economic factors became the only driving force in the food system. The acknowledgement that the socio-economic interactions are indeed one of the fundamental components of any agroecosystems came to light in 1982, with the article Agroecologia del Tropico Americano by Montaldo. The author argues that the socio-economic context cannot be separated from the agricultural systems when designing agricultural practices.

In 1995 Edens et al. in Sustainable Agriculture and Integrated Farming Systems solidified this idea proving his point by devoting special sections to economics of the systems, ecological impacts, and ethics and values in agriculture. Actually, 1985 ended up being a fertile and creative year for the new discipline. For instance in the same year, Miguel Altieri integrated how consolidation of the farms, and cropping systems impact pest populations. In addition, Gliessman highlighted that socio-economic, technological, and ecological components give rise to producer choices of food production systems. These pioneering agroecologists have helped to frame the foundation of what we today consider the interdisciplinary field of agroecology and have led to advances in a number of farming systems. In Asian rice, for example, crop diversification by growing flowering crops in strips beside rice fields has recently been demonstrated to reduce pests so effectively (by the flower nectar attracting and supporting parasitoids and predators) that insecticide spraying is reduced by 70%, yields increase by 5%, together resulting in an economic advantage of 7.5%(Gurr et al., 2016).

By Region

The principles of agroecology are expressed differently depending on local ecological and social contexts.

Latin America

Latin America's experiences with North American Green Revolution agricultural techniques have opened space for agroecologists. Traditional or indigenous knowledge represents a wealth of possibility for agroecologists, including "exchange of wisdoms." See Miguel Alteiri's *Enhancing the Productivity of Latin American Traditional Peasant Farming Systems Through an Agroecological Approach* for information on agroecology in Latin America.

Madagascar

Most of the historical farming in Madagascar has been conducted by indigenous peoples. The French colonial period disturbed a very small percentage of land area, and even included some useful experiments in Sustainable forestry. Slash-and-burn techniques, a component of some shifting cultivation systems have been practised by natives in Madagascar for centuries. As of 2006 some of the major agricultural products from slash-and-burn methods are wood, charcoal and grass for Zebu grazing. These practices have taken perhaps the greatest toll on land fertility since the end of French rule, mainly due to overpopulation pressures.

Agroecosystem

Agroecosystem in Croton-on-Hudson, New York in Westchester County.
Intercropped tomatoes, basil, peppers and eggplants.

An agroecosystem is the basic unit of study in agroecology, and is somewhat arbitrarily defined as a spatially and functionally coherent unit of agricultural activity, and includes the living and non-living components involved in that unit as well as their interactions.

An agroecosystem can be viewed as a subset of a conventional ecosystem. As the name implies, at the core of an agroecosystem lies the human activity of agriculture. However, an agroecosystem is not restricted to the immediate site of agricultural activity (e.g. the farm), but rather includes the region that is impacted by this activity, usually by changes to the complexity of species assemblages and energy flows, as well as to the net nutrient balance. Traditionally an agroecosystem, particularly one managed intensively, is characterized as having a simpler species composition and simpler energy and nutrient flows than "natural" ecosystem. Likewise, agroecosystems are often associated with elevated nutrient input, much of which exits the farm leading to eutrophication of connected ecosystems not directly engaged in agriculture.

The Future for Farming?

Some major organizations are hailing farming within agroecosystems as the way forward for mainstream agriculture. Current farming methods have resulted in over-stretched water resources, high levels of erosion and reduced soil fertility. According to a report by the International Water Management Institute and the United Nations Environment Programme, there is not enough water to continue farming using current practices; therefore how critical water, land, and ecosystem resources are used to boost crop yields must be reconsidered. The report suggested assigning value to ecosystems, recognizing environmental and livelihood tradeoffs, and balancing the rights of a variety of users and interests, as well addressing inequities that sometimes result when such measures are adopted, such as the reallocation of water from poor to rich, the clearing of land to make way for /more productive farmland, or the preservation of a wetland system that limits fishing rights.

A Long Tradition

Forest gardens are probably the world's oldest and most resilient agroecosystem. Forest gardens originated in prehistoric times along jungle-clad river banks and in the wet foothills of monsoon regions. In the gradual process of a family improving their immediate environment, useful tree and vine species were identified, protected and improved whilst undesirable species were eliminated. Eventually superior foreign species were selected and incorporated into the family's garden.

One of the major efforts of disciplines such as agroecology is to promote management styles that blur the distinction between agroecosystems and "natural" ecosystems, both by decreasing the impact of agriculture (increasing the biological and trophic complexity of the agricultural system as well as decreasing the nutrient inputs/outflow) and by increasing awareness that "downstream" effects extend agroecosystems beyond the boundaries of the farm (e.g. the Corn Belt agroecosystem includes the hypoxic zone in the Gulf of Mexico). In the first case, polyculture or buffer strips for wildlife habitat can restore some complexity to a cropping system, while organic farming can reduce nutrient inputs. Efforts of the second type are most common at the watershed scale. An example is the National Association of Conservation Districts' Lake Mendota Watershed Project, which seeks to reduce runoff from the agricultural lands feeding into the lake with the aim of reducing algal blooms.

Agroforestry

Parkland in Burkina Faso: sorghum grown under *Faidherbia albida* and *Borassus akeassii* near Banfora

Agroforestry or agro-sylviculture is a land use management system in which trees or shrubs are grown around or among crops or pastureland. It combines shrubs and trees in agricultural and forestry technologies to create more diverse, productive, profitable, healthy, ecologically sound, and sustainable land-use systems.

As A Science

The theoretical base for agroforestry comes from ecology, via agroecology. From this perspective, agroforestry is one of the three principal land-use sciences. The other two are agriculture and forestry.

Agroforestry has a lot in common with intercropping. Both have two or more plant species (such as nitrogen-fixing plants) in close interaction, both provide multiple outputs, as a consequence, higher overall yields and, because a single application or input is shared, costs are reduced. Beyond these, there are gains specific to agroforestry.

Benefits

Agroforestry systems can be advantageous over conventional agricultural, and forest production methods. They can offer increased productivity, economic benefits, and more diversity in the ecological goods and services provided .(An example of this was seen in trying to conserve Milicia excelsa.)

Biodiversity in agroforestry systems is typically higher than in conventional agricultural systems. With two or more interacting plant species in a given land area, it creates a more complex habitat that can support a wider variety of birds, insects, and other animals. Depending upon the application, impacts of agroforestry can include:

- Reducing poverty through increased production of wood and other tree products for home consumption and sale

- Contributing to food security by restoring the soil fertility for food crops

- Cleaner water through reduced nutrient and soil runoff

- Countering global warming and the risk of hunger by increasing the number of drought-resistant trees and the subsequent production of fruits, nuts and edible oils

- Reducing deforestation and pressure on woodlands by providing farm-grown fuelwood

- Reducing or eliminating the need for toxic chemicals (insecticides, herbicides, etc.)

- Through more diverse farm outputs, improved human nutrition

- In situations where people have limited access to mainstream medicines, providing growing space for medicinal plants

- Increased crop stability

- Multifunctional site use i.e. crop production and animal grazing.

- Typically more drought resistant.

- Stabilises depleted soils from erosion

- Bioremediation

Agroforestry practices may also realize a number of other associated environmental goals, such as:

- Carbon sequestration

- Odour, dust, and noise reduction

- Green space and visual aesthetics

- Enhancement or maintenance of wildlife habitat

Adaptation to Climate Change

There is some evidence that, especially in recent years, poor smallholder farmers are turning to agroforestry as a mean to adapt to the impacts of climate change. A study from the CGIAR research program on Climate Change, Agriculture and Food Security (CCAFS) found from a survey of over 700 households in East Africa that at least 50% of those households had begun planting trees on their farms in a change from their practices 10 years ago. The trees ameliorate the effects of climate change by helping to stabilize erosion, improving water and soil quality and providing yields of fruit, tea, coffee, oil, fodder and medicinal products in addition to their usual harvest. Agroforestry was one of the most widely adopted adaptation strategies in the study, along with the use of improved crop varieties and intercropping.

Applications

Agroforestry represents a wide diversity in application and in practice. One listing includes over 50 distinct uses. The 50 or so applications can be roughly classified under a few broad headings. There are visual similarities between practices in different categories. This is expected as categorization is based around the problems addressed (countering winds, high rainfall, harmful insects, etc.) and the overall economic constraints and objectives (labor and other inputs costs, yield requirements, etc.). The categories include :

- Parklands

- Shade systems

- Crop-over-tree systems

- Alley cropping

- Strip cropping

- Fauna-based systems

- Boundary systems

- Taungyas

- Physical support systems

- Agroforests

- Wind break and shelterbelt.

Parkland

Parklands are visually defined by the presence of trees widely scattered over a large agricultural plot or pasture. The trees are usually of a single species with clear regional favorites. Among the beaks and benefits, the trees offer shade to grazing animals, protect crops against strong wind bursts, provide tree prunings for firewood, and are a roost for insect or rodent-eating birds.

There are other gains. Research with *Faidherbia albida* in Zambia showed that mature trees can sustain maize yields of 4.1 tonnes per hectare compared to 1.3 tonnes per hectare without these trees. Unlike other trees, Faidherbia sheds its nitrogen-rich leaves during the rainy crop growing season so it does not compete with the crop for light, nutrients and water. The leaves then regrow during the dry season and provide land cover and shade for crops.

Shade Systems

With shade applications, crops are purposely raised under tree canopies and within the resulting shady environment. For most uses, the understory crops are shade tolerant or the overstory trees have fairly open canopies. A conspicuous example is shade-grown coffee. This practice reduces weeding costs and improves the quality and taste of the coffee. Just because plants are grown under shade does not necessarily translate into lost or reduced yields. This is because the efficiency of photosynthesis drops off with increasing light intensity, and the rate of photosynthesis hardly increases once the light intensity is over about one tenth that of direct overhead sun. This means that plants under trees can still grow well even though they get less light. By having more than one level of vegetation, it is possible to get more photosynthesis, and overall yields, than with a single canopy layer.

Crop-Over-Tree Systems

Not commonly encountered, crop-over-tree systems employ woody perennials in the role of a cover crop. For this, small shrubs or trees pruned to near ground level are utilized. The purpose, as with any cover crop, is to increase in-soil nutrients and/or to reduce soil erosion.

Alley Cropping

Alley cropping corn fields between rows of walnut trees.

With alley cropping, crop strips alternate with rows of closely spaced tree or hedge species. Normally, the trees are pruned before planting the crop. The cut leafy material is spread over the crop area to provide nutrients for the crop. In addition to nutrients, the hedges serve as windbreaks and eliminate soil erosion.

Alley cropping has been shown to be advantageous in Africa, particularly in relation to improving maize yields in the sub-Saharan region. Use here relies upon the nitrogen fixing tree species *Sesbania sesban, Tephrosia vogelii, Gliricidia sepium* and *Faidherbia albida*. In one example, a ten-year experiment in Malawi showed that, by using the fertilizer tree Gliricidia (*Gliricidia sepium*) on land on which no mineral fertilizer was applied, maize yields averaged 3.3 tonnes per hectare as compared to one tonne per hectare in plots without fertilizer trees nor mineral fertilizers.

Strip Cropping

Strip cropping is similar to alley cropping in that trees alternate with crops. The difference is that, with alley cropping, the trees are in single row. With strip cropping, the trees or shrubs are planted in wide strip. The purpose can be, as with alley cropping, to provide nutrients, in leaf form, to the crop. With strip cropping, the trees can have a purely productive role, providing fruits, nuts, etc. while, at the same time, protecting nearby crops from soil erosion and harmful winds.

Fauna-Based Systems

~ 1970   2004

Silvopasture over the years (Australia).

There are situations where trees benefit fauna. The most common examples are the silvo-pasture where cattle, goats, or sheep browse on grasses grown under trees. In hot climates, the animals are less stressed and put on weight faster when grazing in a cooler, shaded environment. Other variations have these animals directly eating the leaves of trees or shrubs.

There are similar systems for other types of fauna. Deer and hogs gain when living and feeding in a forest ecosystem, especially when the tree forage suits their dietary needs. Another variation, aqua-forestry, is where trees shade fish ponds. In many cases, the fish eat the leaves or fruit from the trees.

Boundary Systems

A riparian buffer bordering a river in Iowa.

There are a number of applications that fall under the heading of a boundary system. These include the living fences, the riparian buffer, and windbreaks.

- A living fence can be a thick hedge or fencing wire strung on living trees. In addition to restricting the movement of people and animals, living fences offer habitat to insect-eating birds and, in the case of a boundary hedge, slow soil erosion.

- Riparian buffers are strips of permanent vegetation located along or near active water-courses or in ditches where water runoff concentrates. The purpose is to keep nutrients and soil from contaminating surface water.

- Windbreaks reduce the velocity of the winds over and around crops. This increases yields through reduced drying of the crop and/or by preventing the crop from toppling in strong wind gusts.

Taungya

Taungya is a vastly used system originating in Burma. In the initial stages of an orchard or tree plantation, the trees are small and widely spaced. The free space between the newly planted trees can accommodate a seasonal crop. Instead of costly weeding, the underutilized area provides an additional output and income. More complex taungyas use the between-tree space for a series of crops. The crops become more shade resistant as the tree canopies grow and the amount of sunlight reaching the ground declines. If a plantation is thinned in the latter stages, this opens further the between-tree cropping opportunities.

Physical Support Systems

In the long history of agriculture, trellises are comparatively recent. Before this, grapes and other vine crops were raised atop pruned trees. Variations of the physical support theme depend upon the type of vine. The advantages come through greater in-field biodiversity. In many cases, the control of weeds, diseases, and insect pests are primary motives.

Agroforests

These are widely found in the humid tropics and are referenced by different names (forest gardening, forest farming, tropical home gardens and, where short-statured trees or shrubs dominate, shrub gardens). Through a complex, diverse mix of trees, shrubs, vines, and seasonal crops, these systems achieve the ecological dynamics of a forest ecosystem. Because of their internal ecology, they tend to be less susceptible to harmful insects, plant diseases, drought, and wind damage.

Historical Use

Agroforestry similar methods were historically utilized by Native Americans. California Indians would prescribe burn oak and other habitats to maintain a 'pyrodiversity collecting model'. This method allowed for greater health of trees and the habitat in general.

Challenges

Agroforestry is relevant to almost all environments and is a potential response to common problems around the globe, and agroforestry systems can be advantageous compared to conventional agriculture or forestry. Yet agroforestry is not very widespread, at least according to current but incomplete USDA surveys as of November, 2013.

As suggested by a survey of extension programs in the United States, some obstacles (ordered most critical to least critical) to agroforestry adoption include:

- Lack of developed markets for products

- Unfamiliarity with technologies

- Lack of awareness of successful agroforestry examples

- Competition between trees, crops, and animals

- Lack of financial assistance

- Lack of apparent profit potential

- Lack of demonstration sites

- Expense of additional management

- Lack of training or expertise

- Lack of knowledge about where to market products

- Lack of technical assistance

- Cannot afford adoption or start up costs, including costs of time

- Unfamiliarity with alternative marketing approaches (e.g. web)

- Unavailability of information about agroforestry

- Apparent inconvenience

- Lack of infrastructure (e.g. buildings, equipment)

- Lack of equipment

- Insufficient land

- Lack of seed/seedling sources

- Lack of scientific research

Some solutions to these obstacles have already been suggested although many depend on particular circumstances which vary from one location to the next.

Conservation Agriculture

Conservation agriculture (CA) can be defined by a statement given by the Food and Agricultural Organization of the United Nations as "a concept for resource-saving agricultural crop production that strives to achieve acceptable profits together with high and sustained production levels while concurrently conserving the environment" (FAO 2007).

Agriculture according to the New Standard Encyclopedia is "one of the most important sectors in the economies of most nations" (New Standard 1992). At the same time conservation is the use of resources in a manner that safely maintains a resource that can be used by humans. Conservation has become critical because the global population has increased over the years and more food needs to be produced every year (New Standard 1992). Sometimes referred to as "agricultural environmental management", conservation agriculture may be sanctioned and funded through conservation programs promulgated through agricultural legislation, such as the U.S. Farm Bill.

Key Principles

The Food and Agricultural Organization of the United Nations (FAO) has determined that CA has three key principles that producers (farmers) can proceed through in the process of CA. These three principles outline what conservationists and producers believe can be done to conserve what we use for a longer period of time.

The first key principle in CA is practicing minimum mechanical soil disturbance which is essential to maintaining minerals within the soil, stopping erosion, and preventing water loss from occurring within the soil. In the past agriculture has looked at soil tillage as a main process in the introduction of new crops to an area. It was believed that tilling the soil would increase fertility within the soil through mineralization that takes place in the soil. Also tilling of soil can cause severe erosion and crusting which leads to a decrease in soil fertility. Today tillage is seen as destroying organic matter that can be found within the soil cover. No-till farming has caught on as a process that can save soil organic levels for a longer period and still allow the soil to be productive for longer periods (FAO 2007). Additionally, the process of tilling can increase time and labor for producing that crop.

When no-till practices are followed, the producer sees a reduction in production cost for a certain crop. Tillage of the ground requires more money in order to fuel tractors or to provide feed for the animals pulling the plough. The producer sees a reduction in labor because he or she does not have to be in the fields as long as a conventional farmer.

The second key principle in CA is much like the first in dealing with protecting the soil. The principle of managing the top soil to create a permanent organic soil cover can allow for growth of organisms within the soil structure. This growth will break down the mulch that is left on the soil surface. The breaking down of this mulch will produce a high organic matter level which will act as a fertilizer for the soil surface. If CA practices were used done for many years and enough organic matter was being built up at the surface, then a layer of mulch would start to form. This layer helps prevent soil erosion from taking place and ruining the soil's profile or layout.

According to the article "The role of conservation agriculture and sustainable agriculture", the layer of mulch that is built up over time will become like a buffer zone between soil and mulch and this will help reduce wind and water erosion. With this comes the protection of the soil's surface when rain falls on the ground. Land that is not protected by a layer of mulch is left open to the elements (Hobbs et al. 2007). This type of ground cover also helps keep the temperature and moisture levels of the soil at a higher level rather than if it was tilled every year (FAO 2007).

The third principle is the practice of crop rotation with more than two species. According to an article published in the *Physiological Transactions of the Royal Society* called "The role of conservation agriculture and sustainable agriculture," crop rotation can be used best as a disease control against other preferred crops (Hobbs et al. 2007). This process will not allow pests such as insects and weeds to be set into a rotation with specific crops. Rotational crops will act as a natural insecticide and herbicide against specific crops. Not allowing insects or weeds to establish a pattern will help to eliminate problems with yield reduction and infestations within fields (FAO 2007). Crop rotation can also help build up soil infrastructure. Establishing crops in a rotation allows for an extensive buildup of rooting zones which will allow for better water infiltration (Hobbs et al. 2007).

Organic molecules in the soil break down into phosphates, nitrates and other beneficial elements which are thus better absorbed by plants. Plowing increases the amount of oxygen in the soil and increases the aerobic processes, hastening the breakdown of organic material. Thus more nutrients are available for the next crop but, at the same time, the soil is depleted more quickly of its nutrient reserves.

Examples

In conservation agriculture there are many examples that can be looked towards as a way of farming and at the same time conserving. These practices are well known by most producers. The process of no-till is one that follows the first principle of CA, causing minimal mechanical soil disturbance. No-till also brings other benefits to the producer . According to the FAO, tillage is one of the most "energy consuming" processes that can be used: It requires a lot of labor, time, and fuel to till. Producers can save 30% to 40% of time and labor by practicing the no-till process. (FAO 3020)

Besides conserving the soil, there are other examples of how CA is used. According to an article in *Science* called "Farming and the Fate of Wild Nature" there are two more kinds of CA . The practice of wildlife-friendly farming and land sparing are ideas for producers who are looking to practice better conservation towards biodiversity (Green, et al. 2005).

Wildlife-Friendly Farming

Wildlife-friendly farming is a practice of setting aside land that will not be developed by the producer (farmer). This land will be set aside so that biodiversity has a chance to establish itself in areas with agricultural fields. At the same time, the producer is attempting to lower the amount of fertilizer and pesticides used on the fields so that organisms and microbial activity have a chance to establish themselves in the soil and habitat (Green, et al. 2005). But as in all systems, not all can be perfect. To create a habitat suitable for biodiversity something has to be reduced, and as in this case for agriculture farmers, yields can be reduced. This is where the second idea of land sparing can be looked on as an alternative.

Land Sparing

Land sparing is another way that producer and conservationist can be on the same page. Land sparing advocates for the land that is being used for agricultural purposes to continue to produce crops at increased yield. With an increase in yield on all land that is in use, other land can be set aside for conservation and production for biodiversity. Agricultural land stays in production but would have to increase its yield potential to keep up with demand. Land that is not being put into agriculture would be used for conserving biodiversity (Green, et al. 2005).

Benefits

In the field of CA there are many benefits that both the producer and conservationist can obtain.

On the side of the conservationist, CA can be seen as beneficial because there is an effort to conserve what people use every day. Since agriculture is one of the most destructive forces against biodiversity, CA can change the way humans produce food and energy. With conservation come environmental benefits of CA. These benefits include less erosion possibilities, better water conservation, improvement in air quality due to lower emissions being produced, and a chance for larger biodiversity in a given area.

On the side of the producer and/or farmer, CA can eventually do all that is done in conventional agriculture, and it can conserve better than conventional agriculture. CA according to Theodor Friedrich, who is a specialist in CA, believes "Farmers like it because it gives them a means of conserving, improving, and making more efficient use of their natural resources" (FAO 2006). Producers will find that the benefits of CA will come later rather than sooner. Since CA takes time to build up enough organic matter and have soils become their own fertilizer, the process does not start to work overnight. But if producers make it through the first few years of production, results will start to become more satisfactory.

CA is shown to have even higher yields and higher outputs than conventional agriculture once it has been established over long periods. Also, a producer has the benefit of knowing that the soil in which his crops are grown is a renewable resource. According to New Standard Encyclopedia, soils are a renewable resource, which means that whatever is taken out of the soil can be put back over time (New Standard 1992). As long as good soil upkeep is maintained, the soil will continue to renew itself. This could be very beneficial to a producer who is practicing CA and is looking to keep soils at a productive level for an extended time.

The farmer and/or producer can use this same land in another way when crops have been harvested. The introduction of grazing livestock to a field that once held crops can be beneficial for the producer and also the field itself. Livestock can be used as a natural fertilizer for a producer's field which will then be beneficial for the producer the next year when crops are planted once again. The practice of grazing livestock using CA helps the farmer who raises crops on that field and the farmer who raises the livestock that graze off that field. Livestock produce compost or manure which are a great help in generating soil fertility (Pawley W.H. 1963). The practices of CA and grazing livestock on a field for many years can allow for better yields in the following years as long as these practices continue to be followed.

The FAO believes that there are three major benefits from CA:

- Within fields that are controlled by CA the producer will see an increase in organic matter.

- Increase in water conservation due to the layer of organic matter and ground cover to help eliminate transportation and access runoff.

- Improvement of soil structure and rooting zone.

Future Development

As in any other business, producers and conservationists are always looking towards the future. In this case CA is a very important process to be looked at for future generation. There are many organizations that have been created to help educate and inform producers and conservationists in the world of CA. These organizations can help to inform, conduct research, and buy land in order to preserve animals and plants (New Standard 1992).

Another way in which CA is looking to the future is through prevention. According to the *European Journal of Agronomy* producers are looking for ways to reduce leaching problems within their fields. These producers are using the same principles within CA, in that they are leaving cover over their fields in order to save fields from erosion and leaching of chemicals (Kirchmann & Thorvaldsson 2000). Processes and studies like this are allowing for a better understanding of how to conserve what we are using and finding ways to put back something that may have been lost before.

In the same journal article is presented another way in which producers and conservationists are looking towards the future. Circulation of plant nutrients can be a vital part for conserving the future. An example of this would be the use of animal manure. This process has been used for quite some time now, but the future is looking towards ways to handle and conserve nutrients within manure for a longer time. But besides animal waste, food and urban waste are also being looked towards as a way to use growth within CA (Kirchmann & Thorvaldsson 2000). Turning these products from waste to being used to grow crops and improve yields is something that would be beneficial for conservationists and producers.

Agri-Environment Schemes

In 1992, 'agri-environment schemes' became compulsory for all European Union Member States. In the following years the main purpose of these schemes changed slightly. Initially, they sought to protect threatened habitats, but gradually shifted their focus to the prevention of the loss of wildlife from agricultural landscapes. Most recently, the schemes are placing more emphasis on improving the services that the land can provide to humans (e.g. pollination). Overall, farmers involved in the scheme aim to practice environmentally friendlier farming techniques such as: reducing the use of pesticides, managing or altering their land to increase more wildlife friendly habitats (e.g. increasing areas of trees and bushes), reducing irrigation, conserving soil, and organic farming. As the changes in practices that ensure the protection of the environment are costly to farmers, the EU developed agri-environment schemes to financially compensate individual farmers for applying these changes and therefore increased the implementation of conservation agriculture. The schemes are voluntary for farmers. Once joined, they commit to a minimum of five years during which they have to adopt various sustainable farming techniques. According to the Euro-stat website, in 2009 the agricultural area enrolled in agri-environment schemes covered 38.5 million hectares (20.9% of agricultural land in the 27 member states of the EU at the time) (Agri-environmental indicator 2015). The European Commission spent a total of €3.23 billion on agri-environment schemes in 2012, significantly exceeding the cost of managing special sites of conservation (Natura 2000) that year, which came to a total of €39.6 million (Batáry et al. 2015).

There are two main types of agri-environment schemes which have shown different outcomes. Out-of-production schemes tend to be used in extensive farming practices (where the farming land is widespread and less intensive farming is practiced), and focus on improving or setting land aside that will not be used for the production of food, for example, the addition of wildflower strips. In-production schemes (used for a smaller scale, but more intensively farmed land) focus on the sustainable management of arable crops or grassland, for example reduction of pesticides, reduction of grassland mowing, and most commonly, organic farming. In a 2015 review of studies examining the effects of the two schemes, it was found that out-of-production schemes had a higher success rate at enhancing the number of thriving species around the land. The reason behind this is thought to be the scheme's focus on enhancing specific species by providing them with more unaltered habitats, which results in more food resources for the specific species. On the other hand, in-production schemes attempt to enhance the quality of the land in general, and are thus less species specific. Based on the findings, the reviewers suggest that schemes which more specifically target the declining groups of species, may be more effective. The findings and the targets will be implemented between 2015 and 2020, so that by 2025, the effectiveness of these schemes can be re-assessed and will have increased significantly (Batáry et al. 2015).

Problems

As much as conservation agriculture can benefit the world, there are some problems that come with it. There are many reasons why conservation agriculture cannot always be a win-win situation.

There are not enough people who can financially turn from conventional farming to conservation. The process of CA takes time; when a producer first becomes a conservationist, the results can be a financial loss to them. CA is based upon establishing an organic layer and producing its own fertilizer and this may take time. It can be many years before a producer will start to see better yields than he/she has had previously. Another financial undertaking is purchasing of new equipment. When starting to use CA, a producer may have to buy new planters or drills in order to produce effectively. These financial tasks are ones that may impact whether or not a producer decides to switch to CA or not.

With the struggle to adapt comes the struggle to make CA grow across the globe. CA has not spread as quickly as most conservationists would like. The reason for this is because there is not enough pressure for producers in places such as North America to change their way of living to a more conservationist outlook. But in the tropics there is more pressure to change to conservation areas because of the limited resources that are available. Places like Europe have also started to catch onto the ideas and principles of CA, but still nothing much is being done to change due to there being a minimal amount of pressure for people to change their ways of living (FAO 2006).

With CA comes the idea of producing enough food. With cutting back in fertilizer, not tilling the ground, and other processes comes the responsibility to feed the world. According to the Population Reference Bureau, there were around 6.08 billion people on Earth in the year 2000. By 2050 there will be an estimated 9.1 billion people. With this increase comes the responsibility for producers to increase food supply using the same or less land than we use today. Problems arise in the fact that if CA farms do not produce as much as conventional farms, this leaves the world with less food for more people.

Environmental Impact of Agriculture

Water pollution in a rural stream due to runoff from farming activity in New Zealand.

The environmental impact of agriculture varies based on the wide variety of agricultural practices employed around the world. Ultimately, the environmental impact depends on the production practices of the system used by farmers. The connection between emissions into the environment and the farming system is indirect, as it also depends on other climate variables such as rainfall and temperature.

There are two types of indicators of environmental impact: "means-based", which is based on the farmer's production methods, and "effect-based", which is the impact that farming methods have on the farming system or on emissions to the environment. An example of a means-based indicator would be the quality of groundwater, that is effected by the amount of nitrogen applied to the soil. An indicator reflecting the loss of nitrate to groundwater would be effect-based.

The environmental impact of agriculture involves a variety of factors from the soil, to water, the air, animal and soil diversity, people, plants, and the food itself. Some of the environmental issues that are related to agriculture are climate change, deforestation, genetic engineering, irrigation problems, pollutants, soil degradation, and waste.

Climate Change

Climate change and agriculture are interrelated processes, both of which take place on a worldwide scale. Global warming is projected to have significant impacts on conditions affecting agriculture, including temperature, precipitation and glacial run-off. These conditions determine the carrying capacity of the biosphere to produce enough food for the human population and domesticated animals. Rising carbon dioxide levels would also have effects, both detrimental and beneficial, on crop yields. Assessment of the effects of global climate changes on agriculture might help to properly anticipate and adapt farming to maximize agricultural production. Although the net impact of climate change on agricultural production is uncertain it is likely that it will shift the suitable growing zones for individual crops. Adjustment to this geographical shift will involve considerable economic costs and social impacts.

At the same time, agriculture has been shown to produce significant effects on climate change, primarily through the production and release of greenhouse gases such as carbon dioxide, methane, and nitrous oxide. In addition, agriculture that practices tillage, fertilization, and pesticide application also releases ammonia, nitrate, phosphorus, and many other pesticides that affect air, water, and soil quality, as well as biodiversity. Agriculture also alters the Earth's land cover, which can change its ability to absorb or reflect heat and light, thus contributing to radiative forcing. Land use change such as deforestation and desertification, together with use of fossil fuels, are the major anthropogenic sources of carbon dioxide; agriculture itself is the major contributor to increasing methane and nitrous oxide concentrations in earth's atmosphere.

Deforestation

Deforestation is clearing the Earth's forests on a large scale worldwide and resulting in many land damages. One of the causes of deforestation is to clear land for pasture or crops. According to British environmentalist Norman Myers, 5% of deforestation is due to cattle ranching, 19% due to over-heavy logging, 22% due to the growing sector of palm oil plantations, and 54% due to slash-and-burn farming.

Deforestation causes the loss of habitat for millions of species, and is also a driver of climate change. Trees act as a carbon sink: that is, they absorb carbon dioxide, an unwanted greenhouse gas, out of the atmosphere. Removing trees releases carbon dioxide into the atmosphere and leaves behind fewer trees to absorb the increasing amount of carbon dioxide in the air. In this way, deforestation exacerbates climate change. When trees are removed from forests, the soils tend to dry out because there is no longer shade, and there are not enough trees to assist in the water cycle by returning water vapor back to the environment. With no trees, landscapes that were once forests can potentially become barren deserts. The removal of trees also causes extreme fluctuations in temperature.

In 2000 the United Nations Food and Agriculture Organization (FAO) found that "the role of population dynamics in a local setting may vary from decisive to negligible," and that deforestation can result from "a combination of population pressure and stagnating economic, social and technological conditions."

Genetic Engineering

Genetically engineered crops are herbicide-tolerant, and their overuse has created herbicide resistant "super weeds", which may ultimately increase the use of herbicides. Seed contamination is another problem of genetic engineering; it can occur from wind or bee pollination that is blown from genetically-engineered crops to normal crops. About 50% of corn and soybean samples and more than 80% of canola samples were found to be contaminated by Monsanto's (genetic engineering company) genes. This accidental contamination can cause organic farmers to lose a lot of money because they need to recall their products. There are various cases of this such as in the corn and alfalfa industry.

Irrigation

Irrigation can lead to a number of problems:

Among some of these problems is the depletion of underground aquifers through overdrafting. Soil can be over-irrigated because of poor distribution uniformity or management wastes water, chemicals, and may lead to water pollution. Over-irrigation can cause deep drainage from rising water tables that can lead to problems of irrigation salinity requiring watertable control by some form of subsurface land drainage. However, if the soil is under irrigated, it gives poor soil salinity control which leads to increased soil salinity with consequent buildup of toxic salts on soil surface in areas with high evaporation. This requires either leaching to remove these salts and a method of drainage to carry the salts away. Irrigation with saline or high-sodium water may damage soil structure owing to the formation of alkaline soil.

Pollutants

Synthetic pesticides are the most widespread method of controlling pests in agriculture. Pesticides can leach through the soil and enter the groundwater, as well as linger in food products and result in death in humans. Pesticides can also kill non-target plants, birds, fish and other wildlife. A wide range of agricultural chemicals are used and some become pollutants through use, misuse, or ignorance. Pollutants from agriculture have a huge effect on water quality. Agricultural nonpoint source (NPS) solution impacts lakes, rivers, wetlands, estuaries, and groundwater. Agricultural NPS can be caused by poorly managed animal feeding operations, overgrazing, plowing, fertilizer, and improper, excessive, or badly timed use of Pesticides. Pollutants from farming include sediments, nutrients, pathogens, pesticides, metals, and salts.

Listed below are additional and specific problems that may arise with the release of pollutants from agriculture.

- Pesticide drift
    - soil contamination
    - air pollution *spray drift*
- Pesticides, especially those based on organochloride
- Pesticide residue in foods

- Pesticide toxicity to bees
    - List of crop plants pollinated by bees
    - Pollination management
- Bioremediation

Soil Degradation

Soil degradation is the decline in soil quality that can be a result of many factors, especially from agriculture. Soils hold the majority of the world's biodiversity, and healthy soils are essential for food production and an adequate water supply. Common attributes of soil degradation can be salting, waterlogging, compaction, pesticide contamination, decline in soil structure quality, loss of fertility, changes in soil acidity, alkalinity, salinity, and erosion. Soil degradation also has a huge impact on biological degradation, which affects the microbial community of the soil and can alter nutrient cycling, pest and disease control, and chemical transformation properties of the soil.

- soil contamination
    - sedimentation

Waste

Plasticulture is the use of plastic mulch in agriculture. Farmers use plastic sheets as mulch to cover 50-70% of the soil and allows them to use drip irrigation systems to have better control over soil nutrients and moisture. Rain is not required in this system, and farms that use plasticulture are built to encourage the fastest runoff of rain. The use of pesticides with plasticulture allows pesticides to be transported easier in the surface runoff towards wetlands or tidal creeks. The runoff from pesticides and chemicals in the plastic can cause serious deformations and death in shellfish as the runoff carries the chemicals towards the oceans.

In addition to the increased runoff that results from plasticulture, there is also the problem of the increased amount of waste form the plastic mulch itself. The use of plastic mulch for vegetables, strawberries, and other row and orchard crops exceeds 110 million pounds annually in the United States. Most plastic ends up in the landfill, although there are other disposal options such as disking mulches into the soil, on-site burying, on-site storage, reuse, recycling, and incineration. The incineration and recycling options are complicated by the variety of the types of plastics that are used and by the geographic dispersal of the plastics. Plastics also contain stabilizers and dyes as well as heavy metals, which limits the amount of products that can be recycled. Research is continually being conducted on creating biodegradable or photodegradable mulches. While there has been minor success with this, there is also the problem of how long the plastic takes to degrade, as many biodegradable products take a long time to break down.

Issues by region

The environmental impact of agriculture can vary depending on the region as well as the type of agriculture production method that is being used. Listed below are some specific environmental issues in a various different regions around the world.

- Hedgerow removal in the United Kingdom.

- Soil salinisation, especially in Australia.

- Phosphate mining in Nauru

- Methane emissions from livestock in New Zealand.

- Environmentalists attribute the hypoxic zone in the Gulf of Mexico as being encouraged by nitrogen fertilization of the algae bloom.

Sustainable Agriculture

The exponential population increase in recent decades has increased the practice of agricultural land conversion to meet demand for food which in turn has increased the effects on the environment. The global population is still increasing and will eventually stabilise, as some critics doubt that food production, due to lower yields from global warming, can support the global population.

Deeply moved by the lives of several farmers in Maharashtra (India), Mrs Mrunalini Bhosale is planning to moulde a Civil Society Origination to support their rehabilitation, through education, bringing in of better farming practices and other beneficiary options. As an initial step for it Mrunalini Bhosale have directed a debut movie Kapus Kondyachi Goshta which goes through the struggle of farmers and a courageous saga of women spirit.

Organic farming is a multifaceted sustainable agriculture set of practices that can have a lower impact on the environment at the small scale. However in most cases organic farming results in lower yields in terms of production per unit area and per unit of irrigation water. Therefore, widespread adoption of organic agriculture will require additional land to be cleared and water resources extracted to meet the same level of production.

Other specific methods include: permaculture; and biodynamic agriculture which incorporates a spiritual element.

- Category: Sustainable agriculture

- Biological pest control

Climate Change and Agriculture

Human greenhouse gas emissions by sector, in the year 2010. "AFOLU" stands for "agriculture, forestry, and other land use".

Climate change and agriculture are interrelated processes, both of which take place on a global scale. Climate change affects agriculture in a number of ways, including through changes in average temperatures, rainfall, and climate extremes (e.g., heat waves); changes in pests and diseases; changes in atmospheric carbon dioxide and ground-level ozone concentrations; changes in the nutritional quality of some foods; and changes in sea level.

Net crop production in selected tropical countries and worldwide (2004-6=100)

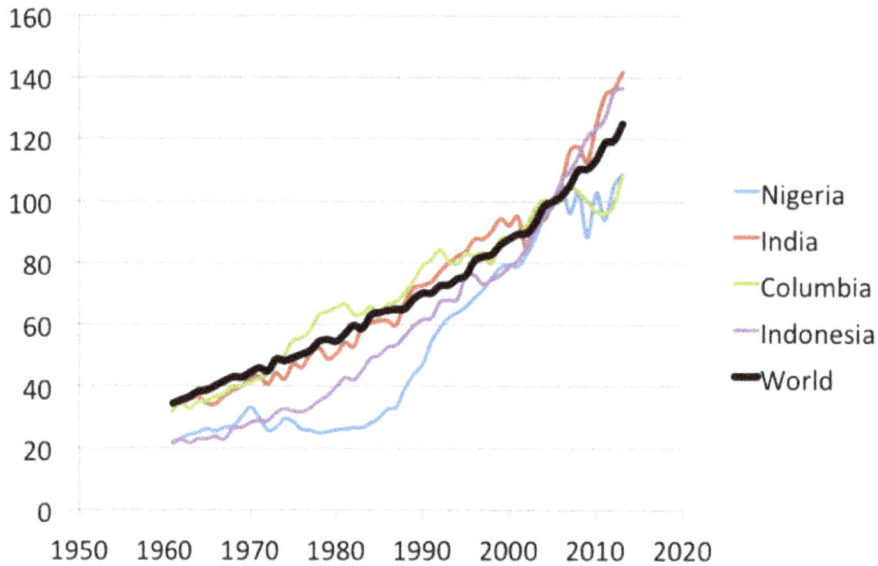

Graph of net crop production worldwide and in selected tropical countries. Raw data from the United Nations.

Climate change is already affecting agriculture, with effects unevenly distributed across the world. Future climate change will likely negatively affect crop production in low latitude countries, while effects in northern latitudes may be positive or negative. Climate change will probably increase the risk of food insecurity for some vulnerable groups, such as the poor.

Agriculture contributes to climate change by (1) anthropogenic emissions of greenhouse gases (GHGs), and (2) by the conversion of non-agricultural land (e.g., forests) into agricultural land. Agriculture, forestry and land-use change contributed around 20 to 25% to global annual emissions in 2010.

There are range of policies that can reduce the risk of negative climate change impacts on agriculture, and to reduce GHG emissions from the agriculture sector.

Impact of Climate Change on Agriculture

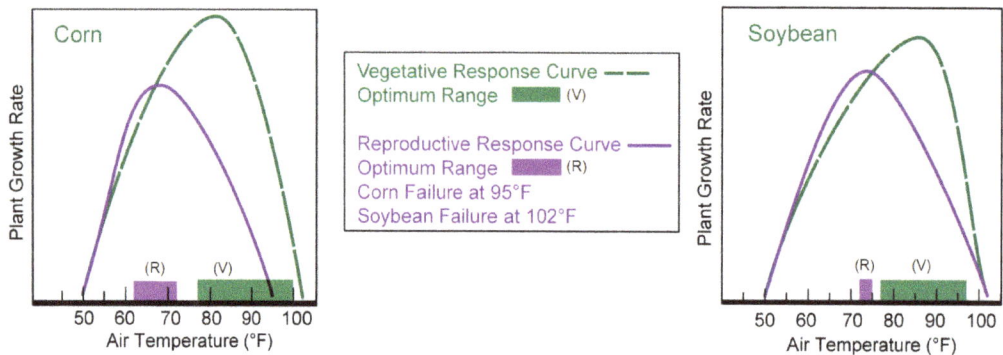

ARS USDA

For each plant variety, there is an optimal temperature for vegetative growth, with growth dropping off as temperatures increase or decrease. Similarly, there is a range of temperatures at which a plant will produce seed. Outside of this range, the plant will not reproduce. As the graphs show, corn will fail to reproduce at temperatures above 95 °F and soybean above 102 °F.

Despite technological advances, such as improved varieties, genetically modified organisms, and irrigation systems, weather is still a key factor in agricultural productivity, as well as soil properties and natural communities. The effect of climate on agriculture is related to variabilities in local climates rather than in global climate patterns. The Earth's average surface temperature has increased by 1.5 °F (0.83 °C) since 1880. Consequently, agronomists consider any assessment has to be individually consider each local area.

On the other hand, agricultural trade has grown in recent years, and now provides significant amounts of food, on a national level to major importing countries, as well as comfortable income to exporting ones. The international aspect of trade and security in terms of food implies the need to also consider the effects of climate change on a global scale.

A 2008 study published in *Science* suggested that, due to climate change, "southern Africa could lose more than 30% of its main crop, maize, by 2030. In South Asia losses of many regional staples, such as rice, millet and maize could top 10%".

The Intergovernmental Panel on Climate Change (IPCC) has produced several reports that have assessed the scientific literature on climate change. The IPCC Third Assessment Report, published in 2001, concluded that the poorest countries would be hardest hit, with reductions in crop yields in most tropical and sub-tropical regions due to decreased water availability, and new or changed insect pest incidence. In Africa and Latin America many rainfed crops are near their maximum temperature tolerance, so that yields are likely to fall sharply for even small climate changes; falls in agricultural productivity of up to 30% over the 21st century are projected. Marine life and the fishing industry will also be severely affected in some places.

Climate change induced by increasing greenhouse gases is likely to affect crops differently from region to region. For example, average crop yield is expected to drop down to 50% in Pakistan according to the UKMO scenario whereas corn production in Europe is expected to grow up to 25% in optimum hydrologic conditions.

More favourable effects on yield tend to depend to a large extent on realization of the potentially beneficial effects of carbon dioxide on crop growth and increase of efficiency in water use. Decrease in potential yields is likely to be caused by shortening of the growing period, decrease in water availability and poor vernalization.

In the long run, the climatic change could affect agriculture in several ways :

- *productivity*, in terms of quantity and quality of crops

- *agricultural practices*, through changes of water use (irrigation) and agricultural inputs such as herbicides, insecticides and fertilizers

- *environmental effects*, in particular in relation of frequency and intensity of soil drainage (leading to nitrogen leaching), soil erosion, reduction of crop diversity

- *rural space*, through the loss and gain of cultivated lands, land speculation, land renunciation, and hydraulic amenities.

- *adaptation*, organisms may become more or less competitive, as well as humans may develop urgency to develop more competitive organisms, such as flood resistant or salt resistant varieties of rice.

They are large uncertainties to uncover, particularly because there is lack of information on many specific local regions, and include the uncertainties on magnitude of climate change, the effects of technological changes on productivity, global food demands, and the numerous possibilities of adaptation.

Most agronomists believe that agricultural production will be mostly affected by the severity and pace of climate change, not so much by gradual trends in climate. If change is gradual, there may be enough time for biota adjustment. Rapid climate change, however, could harm agriculture in many countries, especially those that are already suffering from rather poor soil and climate conditions, because there is less time for optimum natural selection and adaption.

But much remains unknown about exactly how climate change may affect farming and food security, in part because the role of farmer behaviour is poorly captured by crop-climate models. For instance, Evan Fraser, a geographer at the University of Guelph in Ontario Canada, has conducted a number of studies that show that the socio-economic context of farming may play a huge role in determining whether a drought has a major, or an insignificant impact on crop production. In some cases, it seems that even minor droughts have big impacts on food security (such as what happened in Ethiopia in the early 1980s where a minor drought triggered a massive famine), versus cases where even relatively large weather-related problems were adapted to without much hardship. Evan Fraser combines socio-economic models along with climatic models to identify "vulnerability hotspots" One such study has identified US maize (corn) production as particularly vulnerable to climate change because it is expected to be exposed to worse droughts, but it does not have the socio-economic conditions that suggest farmers will adapt to these changing conditions.

Observed Impacts

So far, the effects of regional climate change on agriculture have been relatively limited. Changes in crop phenology provide important evidence of the response to recent regional climate change. Phenology is the study of natural phenomena that recur periodically, and how these phenomena relate to climate and seasonal changes. A significant advance in phenology has been observed for agriculture and forestry in large parts of the Northern Hemisphere.

Droughts have been occurring more frequently because of global warming and they are expected to become more frequent and intense in Africa, southern Europe, the Middle East, most of the Americas, Australia, and Southeast Asia. Their impacts are aggravated because of increased water demand, population growth, urban expansion, and environmental protection efforts in many areas. Droughts result in crop failures and the loss of pasture grazing land for livestock.

Examples

As of the decade starting in 2010, many hot countries have thriving agricultural sectors.

Banana farm at Chinawal village in Jalgaon district, India

Jalgaon district, India, has an average temperature which ranges from 20.2C in December to 29.8C in May, and an average precipitation of 750mm/year. It produces bananas at a rate that would make it the world's seventh-largest banana producer if it were a country.

During the period 1990-2012, Nigeria had an average temperature which ranged from a low of 24.9C in January to a high of 30.4C in April. According to the Food and Agriculture Organization of the United Nations (FAO), Nigeria is by far the world's largest producer of yams, producing over 38 million tonnes in 2012. The second through 8th largest yam producers were all nearby African countries, with the largest non-African producer, Papua New Guinea, producing less than 1% of Nigerian production.

In 2013, according to the FAO, Brazil and India were by far the world's leading producers of Sugarcane, with a combined production of over 1 billion tonnes, or over half of worldwide production.

Projections

As part of the IPCC's Fourth Assessment Report, Schneider *et al.* (2007) projected the potential future effects of climate change on agriculture. With low to medium confidence, they concluded that for about a 1 to 3 °C global mean temperature increase (by 2100, relative to the 1990–2000 average level) there would be productivity decreases for some cereals in low latitudes, and productivity increases in high latitudes. In the IPCC Fourth Assessment Report, "low confidence" means that a particular finding has about a 2 out of 10 chance of being correct, based on expert judgement. "Medium confidence" has about a 5 out of 10 chance of being correct. Over the same time period, with medium confidence, global production potential was projected to:

- increase up to around 3 °C,
- very likely decrease above about 3 °C.

Most of the studies on global agriculture assessed by Schneider *et al.* (2007) had not incorporated a number of critical factors, including changes in extreme events, or the spread of pests and diseases. Studies had also not considered the development of specific practices or technologies to aid adaptation to climate change.

The US National Research Council (US NRC, 2011) assessed the literature on the effects of climate change on crop yields. US NRC (2011) stressed the uncertainties in their projections of changes in crop yields.

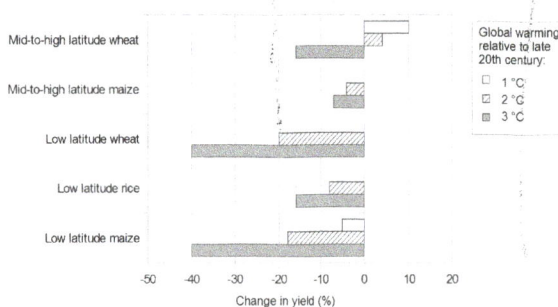

Projected changes in crop yields at different latitudes with global warming. This graph is based on several studies.

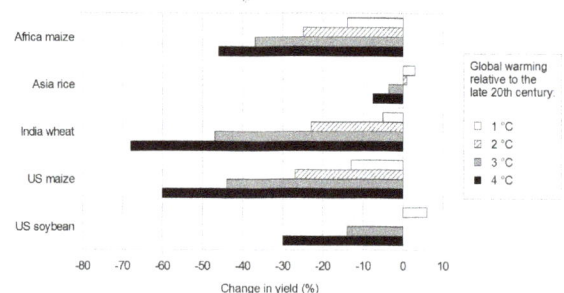

Projected changes in yields of selected crops with global warming. This graph is based on several studies.

Their central estimates of changes in crop yields are shown above. Actual changes in yields may be above or below these central estimates. US NRC (2011) also provided an estimated the "likely" range of changes in yields. "Likely" means a greater than 67% chance of being correct, based on expert judgement. The likely ranges are summarized in the image descriptions of the two graphs.

Food Security

The IPCC Fourth Assessment Report also describes the impact of climate change on food security. Projections suggested that there could be large decreases in hunger globally by 2080, compared to the (then-current) 2006 level. Reductions in hunger were driven by projected social and economic development. For reference, the Food and Agriculture Organization has estimated that in 2006, the number of people undernourished globally was 820 million. Three scenarios *without* climate change (SRES A1, B1, B2) projected 100-130 million undernourished by the year 2080, while another scenario without climate change (SRES A2) projected 770 million undernourished. Based on an expert assessment of all of the evidence, these projections were thought to have about a 5-in-10 chance of being correct.

The same set of greenhouse gas and socio-economic scenarios were also used in projections that included the effects of climate change. *Including* climate change, three scenarios (SRES A1, B1, B2) projected 100-380 million undernourished by the year 2080, while another scenario with climate change (SRES A2) projected 740-1,300 million undernourished. These projections were thought to have between a 2-in-10 and 5-in-10 chance of being correct.

Projections also suggested regional changes in the global distribution of hunger. By 2080, sub-Saharan Africa may overtake Asia as the world's most food-insecure region. This is mainly due to projected social and economic changes, rather than climate change.

Individual Studies

Cline (2008) looked at how climate change might affect agricultural productivity in the 2080s. His study assumes that no efforts are made to reduce anthropogenic greenhouse gas emissions, leading to global warming of 3.3 °C above the pre-industrial level. He concluded that global agricultural productivity could be negatively affected by climate change, with the worst effects in developing countries.

Lobell *et al.* (2008a) assessed how climate change might affect 12 food-insecure regions in 2030. The purpose of their analysis was to assess where adaptation measures to climate change should be prioritized. They found that without sufficient adaptation measures, South Asia and South Africa would likely suffer negative impacts on several crops which are important to large food insecure human populations.

Battisti and Naylor (2009) looked at how increased seasonal temperatures might affect agricultural productivity. Projections by the IPCC suggest that with climate change, high seasonal temperatures will become widespread, with the likelihood of extreme temperatures increasing through the second half of the 21st century. Battisti and Naylor (2009) concluded that such changes could have very serious effects on agriculture, particularly in the tropics. They suggest that major, near-term, investments in adaptation measures could reduce these risks.

"Climate change merely increases the urgency of reforming trade policies to ensure that global food security needs are met" said C. Bellmann, ICTSD Programmes Director. A 2009 ICTSD-IPC study by Jodie Keane suggests that climate change could cause farm output in sub-Saharan Africa to decrease by 12 percent by 2080 - although in some African countries this figure could be as much as 60 percent, with agricultural exports declining by up to one fifth in others.

Adapting to climate change could cost the agriculture sector $14bn globally a year, the study finds.

Regional

Africa

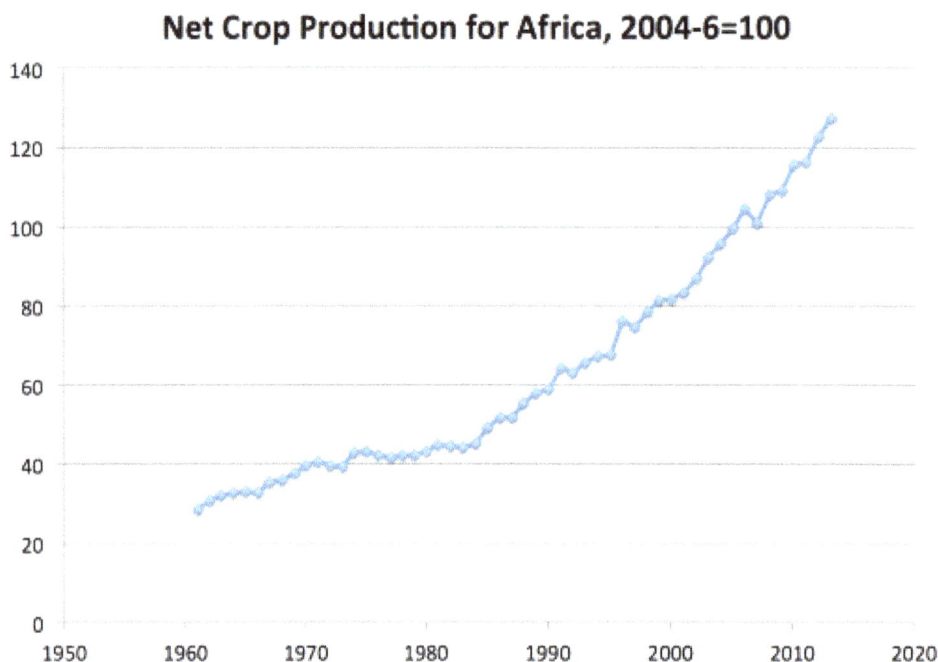

Net Crop Production for Africa, 2004-6=100

African crop production. Raw data from the United Nations.[6]

In Africa, IPCC (2007:13) projected that climate variability and change would severely compromise agricultural production and access to food. This projection was assigned "high confidence."

Africa's geography makes it particularly vulnerable to climate change, and seventy per cent of the population rely on rain-fed agriculture for their livelihoods. Tanzania's official report on climate change suggests that the areas that usually get two rainfalls in the year will probably get more, and those that get only one rainy season will get far less. As of 2005, the net result was expected to be that 33% less maize—the country's staple crop—would be grown.

Asia

In East and Southeast Asia, IPCC (2007:13) projected that crop yields could increase up to 20% by the mid-21st century. In Central and South Asia, projections suggested that yields might decrease by up to 30%, over the same time period. These projections were assigned "medium confidence." Taken together, the risk of hunger was projected to remain very high in several developing countries.

More detailed analysis of rice yields by the International Rice Research Institute forecast 20% reduction in yields over the region per degree Celsius of temperature rise. Rice becomes sterile if exposed to temperatures above 35 degrees for more than one hour during flowering and consequently produces no grain.

A 2013 study by the International Crops Research Institute for the Semi-Arid Tropics (ICRISAT) aimed to find science-based, pro-poor approaches and techniques that would enable Asia's agricultural systems to cope with climate change, while benefitting poor and vulnerable farmers. The study's recommendations ranged from improving the use of climate information in local planning and strengthening weather-based agro-advisory services, to stimulating diversification of rural household incomes and providing incentives to farmers to adopt natural resource conservation measures to enhance forest cover, replenish groundwater and use renewable energy. A 2014 study found that warming had increased maize yields in the Heilongjiang region of China had increased by between 7 and 17% per decade as a result of rising temperatures.

Australia and New Zealand

Hennessy *et al..* (2007:509) assessed the literature for Australia and New Zealand. They concluded that without further adaptation to climate change, projected impacts would likely be substantial: By 2030, production from agriculture and forestry was projected to decline over much of southern and eastern Australia, and over parts of eastern New Zealand; In New Zealand, initial benefits were projected close to major rivers and in western and southern areas. Hennessy *et al..* (2007:509) placed high confidence in these projections.

Europe

With high confidence, IPCC (2007:14) projected that in Southern Europe, climate change would reduce crop productivity. In Central and Eastern Europe, forest productivity was expected to decline. In Northern Europe, the initial effect of climate change was projected to increase crop yields.

Latin America

The major agricultural products of Latin American regions include livestock and grains, such as maize, wheat, soybeans, and rice. Increased temperatures and altered hydrological cycles are predicted to translate to shorter growing seasons, overall reduced biomass production, and lower grain yields. Brazil, Mexico and Argentina alone contribute 70-90% of the total agricultural production in Latin America. In these and other dry regions, maize production is expected to decrease. A study summarizing a number of impact studies of climate change on agriculture in Latin America indicated that wheat is expected to decrease in Brazil, Argentina and Uruguay. Livestock, which is the main agricultural product for parts of Argentina, Uruguay, southern Brazil, Venezuela, and Colombia is likely to be reduced. Variability in the degree of production decrease among different regions of Latin America is likely. For example, one 2003 study that estimated future maize production in Latin America predicted that by 2055 maize in eastern Brazil will have moderate changes while Venezuela is expected to have drastic decreases.

Suggested potential adaptation strategies to mitigate the impacts of global warming on agriculture in Latin America include using plant breeding technologies and installing irrigation infrastructure.

Climate Justice and Subsistence Farmers in Latin America

Several studies that investigated the impacts of climate change on agriculture in Latin America suggest that in the poorer countries of Latin America, agriculture composes the most important economic sector and the primary form of sustenance for small farmers. Maize is the only grain still produced as a sustenance crop on small farms in Latin American nations. Scholars argue that the projected decrease of this grain and other crops will threaten the welfare and the economic development of subsistence communities in Latin America. Food security is of particular concern to rural areas that have weak or non-existent food markets to rely on in the case food shortages.

According to scholars who considered the environmental justice implications of climate change, the expected impacts of climate change on subsistence farmers in Latin America and other developing regions are unjust for two reasons. First, subsistence farmers in developing countries, including those in Latin America are disproportionately vulnerable to climate change Second, these nations were the least responsible for causing the problem of anthropogenic induced climate.

According to researchers John F. Morton and T. Roberts, disproportionate vulnerability to climate disasters is socially determined. For example, socioeconomic and policy trends affecting smallholder and subsistence farmers limit their capacity to adapt to change. According to W. Baethgen who studied the vulnerability of Latin American agriculture to climate change, a history of policies and economic dynamics has negatively impacted rural farmers. During the 1950s and through the 1980s, high inflation and appreciated real exchange rates reduced the value of agricultural exports. As a result, farmers in Latin America received lower prices for their products compared to world market prices. Following these outcomes, Latin American policies and national crop programs aimed to stimulate agricultural intensification. These national crop programs benefitted larger commercial farmers more. In the 1980s and 1990s low world market prices for cereals and livestock resulted in decreased agricultural growth and increased rural poverty.

In the book, Fairness in Adaptation to Climate Change, the authors describe the global injustice of climate change between the rich nations of the north, who are the most responsible for global warming and the southern poor countries and minority populations within those countries who are most vulnerable to climate change impacts.

Adaptive planning is challenged by the difficulty of predicting local scale climate change impacts. An expert that considered opportunities for climate change adaptation for rural communities argues that a crucial component to adaptation should include government efforts to lessen the effects of food shortages and famines. This researcher also claims that planning for equitable adaptation and agricultural sustainability will require the engagement of farmers in decision making processes.

North America

A number of studies have been produced which assess the impacts of climate change on agriculture in North America. The IPCC Fourth Assessment Report of agricultural impacts in the region cites 26 different studies. With high confidence, IPCC (2007:14–15) projected that over the first few decades of this century, moderate climate change would increase aggregate yields of rainfed agriculture by 5–20%, but with important variability among regions. Major challenges were projected for crops that are near the warm end of their suitable range or which depend on highly utilized water resources.

Droughts are becoming more frequent and intense in arid and semiarid western North America as temperatures have been rising, advancing the timing and magnitude of spring snow melt floods and reducing river flow volume in summer. Direct effects of climate change include increased heat and water stress, altered crop phenology, and disrupted symbiotic interactions. These effects may be exacerbated by climate changes in river flow, and the combined effects are likely to reduce the abundance of native trees in favor of non-native herbaceous and drought-tolerant competitors, reduce the habitat quality for many native animals, and slow litter decomposition and nutrient cycling. Climate change effects on human water demand and irrigation may intensify these effects.

United States

The US Global Change Research Program (2009) assessed the literature on the impacts of climate change on agriculture in the United States:

- Many crops will benefit from increased atmospheric CO_2 concentrations and low levels of warming, but higher levels of warming will negatively affect growth and yields. Extreme events will likely reduce crop yields.

- Weeds. diseases and insect pests benefit from warming, and will require more attention in regards to pest and weed control.

- Increasing CO_2 concentrations will reduce the land's ability to supply adequate livestock feed. Increased heat, disease, and weather extremes will likely reduce livestock productivity.

According to a paper by Deschenes and Greenstone (2006), predicted increases in temperature and precipitation will have virtually no effect on the most important crops in the US.

Polar Regions (Arctic and Antarctic)

Anisimov et al.. (2007:655) assessed the literature for the polar region (Arctic and Antarctica). With medium confidence, they concluded that the benefits of a less severe climate were dependent on local conditions. One of these benefits was judged to be increased agricultural and forestry opportunities.

For the *Guardian* newspaper, Brown (2005) reported on how climate change had affected agriculture in Iceland. Rising temperatures had made the widespread sowing of barley possible, which had been untenable twenty years ago. Some of the warming was due to a local (possibly temporary) effect via ocean currents from the Caribbean, which had also affected fish stocks.

Small Islands

In a literature assessment, Mimura *et al.* (2007:689) concluded that on small islands, subsistence and commercial agriculture would very likely be adversely affected by climate change. This projection was assigned "high confidence."

Poverty Impacts

Researchers at the Overseas Development Institute (ODI) have investigated the potential impacts climate change could have on agriculture, and how this would affect attempts at alleviating poverty in the developing world. They argued that the effects from moderate climate change are likely to be mixed for developing countries. However, the vulnerability of the poor in developing countries to short term impacts from climate change, notably the increased frequency and severity of adverse weather events is likely to have a negative impact. This, they say, should be taken into account when defining agricultural policy.

Mitigation and Adaptation in Developing Countries

The Intergovernmental Panel on Climate Change (IPCC) has reported that agriculture is responsible for over a quarter of total global greenhouse gas emissions. Given that agriculture's share in global gross domestic product (GDP) is about 4 percent, these figures suggest that agriculture is highly greenhouse gas intensive. Innovative agricultural practices and technologies can play a role in climate mitigation and adaptation. This adaptation and mitigation potential is nowhere more pronounced than in developing countries where agricultural productivity remains low; poverty, vulnerability and food insecurity remain high; and the direct effects of climate change are expected to be especially harsh. Creating the necessary agricultural technologies and harnessing them to enable developing countries to adapt their agricultural systems to changing climate will require innovations in policy and institutions as well. In this context, institutions and policies are important at multiple scales.

Travis Lybbert and Daniel Sumner suggest six policy principles: (1) The best policy and institutional responses will enhance information flows, incentives and flexibility. (2) Policies and institutions that promote economic development and reduce poverty will often improve agricultural adaptation and may also pave the way for more effective climate change mitigation through agriculture. (3) Business as usual among the world's poor is not adequate. (4) Existing technology options must be made more available and accessible without overlooking complementary capacity and investments. (5) Adaptation and mitigation in agriculture will require local responses, but effective policy responses must also reflect global impacts and inter-linkages. (6) Trade will play a critical role in both mitigation and adaptation, but will itself be shaped importantly by climate change.

The Agricultural Model Intercomparison and Improvement Project (AgMIP) was developed in 2010 to evaluate agricultural models and intercompare their ability to predict climate impacts. In sub-Saharan Africa and South Asia, South America and East Asia, AgMIP regional research teams (RRTs) are conducting integrated assessments to improve understanding of agricultural impacts of climate change (including biophysical and economic impacts) at national and regional scales. Other AgMIP initiatives include global gridded modeling, data and information technology (IT) tool development, simulation of crop pests and diseases, site-based crop-climate sensitivity studies, and aggregation and scaling.

Crop Development Models

Models for climate behavior are frequently inconclusive. In order to further study effects of global warming on agriculture, other types of models, such as *crop development models, yield prediction*, quantities of *water or fertilizer consumed*, can be used. Such models condense the knowledge accumulated of the climate, soil, and effects observed of the results of various agricultural practices. They thus could make it possible to test strategies of adaptation to modifications of the environment.

Because these models are necessarily simplifying natural conditions (often based on the assumption that weeds, disease and insect pests are controlled), it is not clear whether the results they give will have an *in-field* reality. However, some results are partly validated with an increasing number of experimental results.

Other models, such as *insect and disease development* models based on climate projections are also used (for example simulation of aphid reproduction or septoria (cereal fungal disease) development).

Scenarios are used in order to estimate climate changes effects on crop development and yield. Each scenario is defined as a set of meteorological variables, based on generally accepted projections. For example, many models are running simulations based on doubled carbon dioxide projections, temperatures raise ranging from 1 °C up to 5 °C, and with rainfall levels an increase or decrease of 20%. Other parameters may include humidity, wind, and solar activity. Scenarios of crop models are testing farm-level adaptation, such as sowing date shift, climate adapted species (vernalisation need, heat and cold resistance), irrigation and fertilizer adaptation, resistance to disease. Most developed models are about wheat, maize, rice and soybean.

Temperature Potential Effect on Growing Period

Duration of crop growth cycles are above all, related to temperature. An increase in temperature will speed up development. In the case of an annual crop, the duration between sowing and harvesting will shorten (for example, the duration in order to harvest corn could shorten between one and four weeks). The shortening of such a cycle could have an adverse effect on productivity because senescence would occur sooner.

Effect of Elevated Carbon Dioxide on Crops

Carbon dioxide is essential to plant growth. Rising CO_2 concentration in the atmosphere can have both positive and negative consequences.

Increased CO_2 is expected to have positive physiological effects by increasing the rate of photosynthesis. This is known as 'carbon dioxide fertilisation'. Currently, the amount of carbon dioxide in the atmosphere is 380 parts per million. In comparison, the amount of oxygen is 210,000 ppm. This means that often plants may be starved of carbon dioxide as the enzyme that fixes CO_2, RuBisCo, also fixes oxygen in the process of photorespiration. The effects of an increase in carbon dioxide would be higher on C3 crops (such as wheat) than on C4 crops (such as maize), because the former is more susceptible to carbon dioxide shortage. Studies have shown that increased CO_2 leads to fewer stomata developing on plants which leads to reduced water usage. Under optimum conditions of temperature and humidity, the yield increase could reach 36%, if the levels of carbon dioxide are doubled. A study in 2014 posited that CO_2 fertilisation is underestimated due to not explicitly representing CO_2 diffusion inside leaves.

Further, few studies have looked at the impact of elevated carbon dioxide concentrations on whole farming systems. Most models study the relationship between CO_2 and productivity in isolation from other factors associated with climate change, such as an increased frequency of extreme weather events, seasonal shifts, and so on.

In 2005, the Royal Society in London concluded that the purported benefits of elevated carbon dioxide concentrations are "likely to be far lower than previously estimated when factors such as increasing ground-level ozone are taken into account."

Effect on Quality

According to the IPCC's TAR, "The importance of climate change impacts on grain and forage quality emerges from new research. For rice, the amylose content of the grain—a major determinant of cooking quality—is increased under elevated CO_2" (Conroy et al., 1994). Cooked rice grain from plants grown in high-CO_2 environments would be firmer than that from today's plants. However, concentrations of iron and zinc, which are important for human nutrition, would be lower (Seneweera and Conroy, 1997). Moreover, the protein content of the grain decreases under combined increases of temperature and CO_2 (Ziska et al., 1997). Studies using FACE have shown that increases in CO_2 lead to decreased concentrations of micronutrients in crop plants. This may have knock-on effects on other parts of ecosystems as herbivores will need to eat more food to gain the same amount of protein.

Studies have shown that higher CO_2 levels lead to reduced plant uptake of nitrogen (and a smaller number showing the same for trace elements such as zinc) resulting in crops with lower nutritional value. This would primarily impact on populations in poorer countries less able to compensate by eating more food, more varied diets, or possibly taking supplements.

Reduced nitrogen content in grazing plants has also been shown to reduce animal productivity in sheep, which depend on microbes in their gut to digest plants, which in turn depend on nitrogen intake.

Agricultural Surfaces and Climate Changes

Climate change may increase the amount of arable land in high-latitude region by reduction of the amount of frozen lands. A 2005 study reports that temperature in Siberia has increased three degree Celsius in average since 1960 (much more than the rest of the world). However, reports about the impact of global warming on Russian agriculture indicate conflicting probable effects : while they expect a northward extension of farmable lands, they also warn of possible productivity losses and increased risk of drought.

Sea levels are expected to get up to one meter higher by 2100, though this projection is disputed. A rise in the sea level would result in an agricultural land loss, in particular in areas such as South East Asia. Erosion, submergence of shorelines, salinity of the water table due to the increased sea levels, could mainly affect agriculture through inundation of low-lying lands.

Low-lying areas such as Bangladesh, India and Vietnam will experience major loss of rice crop if sea levels rise as expected by the end of the century. Vietnam for example relies heavily on its southern tip, where the Mekong Delta lies, for rice planting. Any rise in sea level of no more than a meter will drown several km² of rice paddies, rendering Vietnam incapable of producing its main staple and export of rice.

Erosion and Fertility

The warmer atmospheric temperatures observed over the past decades are expected to lead to a more vigorous hydrological cycle, including more extreme rainfall events. Erosion and soil degradation is more likely to occur. Soil fertility would also be affected by global warming. However, because the ratio of soil organic carbon to nitrogen is mediated by soil biology such that it maintains a narrow range, a doubling of soil organic carbon is likely to imply a doubling in the storage of nitrogen in soils as organic nitrogen, thus providing higher available nutrient levels for plants, supporting higher yield potential. The demand for imported fertilizer nitrogen could decrease, and provide the opportunity for changing costly fertilisation strategies.

Due to the extremes of climate that would result, the increase in precipitations would probably result in greater risks of erosion, whilst at the same time providing soil with better hydration, according to the intensity of the rain. The possible evolution of the organic matter in the soil is a highly contested issue: while the increase in the temperature would induce a greater rate in the production of minerals, lessening the soil organic matter content, the atmospheric CO_2 concentration would tend to increase it.

Potential Effects of Global Climate Change on Pests, Diseases and Weeds

A very important point to consider is that weeds would undergo the same acceleration of cycle as cultivated crops, and would also benefit from carbonaceous fertilization. Since most weeds are C3 plants, they are likely to compete even more than now against C4 crops such as corn. However, on the other hand, some results make it possible to think that weedkillers could gain in effectiveness with the temperature increase.

Global warming would cause an increase in rainfall in some areas, which would lead to an increase of atmospheric humidity and the duration of the wet seasons. Combined with higher temperatures, these could favor the development of fungal diseases. Similarly, because of higher temperatures and humidity, there could be an increased pressure from insects and disease vectors.

Glacier Retreat and Disappearance

The continued retreat of glaciers will have a number of different quantitative impacts. In the areas that are heavily dependent on water runoff from glaciers that melt during the warmer summer months, a continuation of the current retreat will eventually deplete the glacial ice and substantially reduce or eliminate runoff. A reduction in runoff will affect the ability to irrigate crops and will reduce summer stream flows necessary to keep dams and reservoirs replenished.

Approximately 2.4 billion people live in the drainage basin of the Himalayan rivers. India, China, Pakistan, Afghanistan, Bangladesh, Nepal and Myanmar could experience floods followed by severe droughts in coming decades. In India alone, the Ganges provides water for drinking and farming for more than 500 million people. The west coast of North America, which gets much of its water from glaciers in mountain ranges such as the Rocky Mountains and Sierra Nevada, also would be affected.

Ozone and UV-B

Some scientists think agriculture could be affected by any decrease in stratospheric ozone, which could increase biologically dangerous ultraviolet radiation B. Excess ultraviolet radiation B can directly affect plant physiology and cause massive amounts of mutations, and indirectly through changed pollinator behavior, though such changes are not simple to quantify. However, it has not yet been ascertained whether an increase in greenhouse gases would decrease stratospheric ozone levels.

In addition, a possible effect of rising temperatures is significantly higher levels of ground-level ozone, which would substantially lower yields.

ENSO Effects on Agriculture

ENSO (El Niño Southern Oscillation) will affect monsoon patterns more intensely in the future as climate change warms up the ocean's water. Crops that lie on the equatorial belt or under the tropical Walker circulation, such as rice, will be affected by varying monsoon patterns and more unpredictable weather. Scheduled planting and harvesting based on weather patterns will become less effective.

Areas such as Indonesia where the main crop consists of rice will be more vulnerable to the increased intensity of ENSO effects in the future of climate change. University of Washington professor, David Battisti, researched the effects of future ENSO patterns on the Indonesian rice agriculture using [IPCC]'s 2007 annual report and 20 different logistical models mapping out climate factors such as wind pressure, sea-level, and humidity, and found that rice harvest will experience a decrease in yield. Bali and Java, which holds 55% of the rice yields in Indonesia, will be likely to experience 9–10% probably of delayed monsoon patterns, which prolongs the hungry season. Normal planting of rice crops begin in October and harevest by January. However, as climate change affects ENSO and consequently delays planting, harvesting will be late and in drier conditions, resulting in less potential yields.

Impact of Agriculture on Climate Change

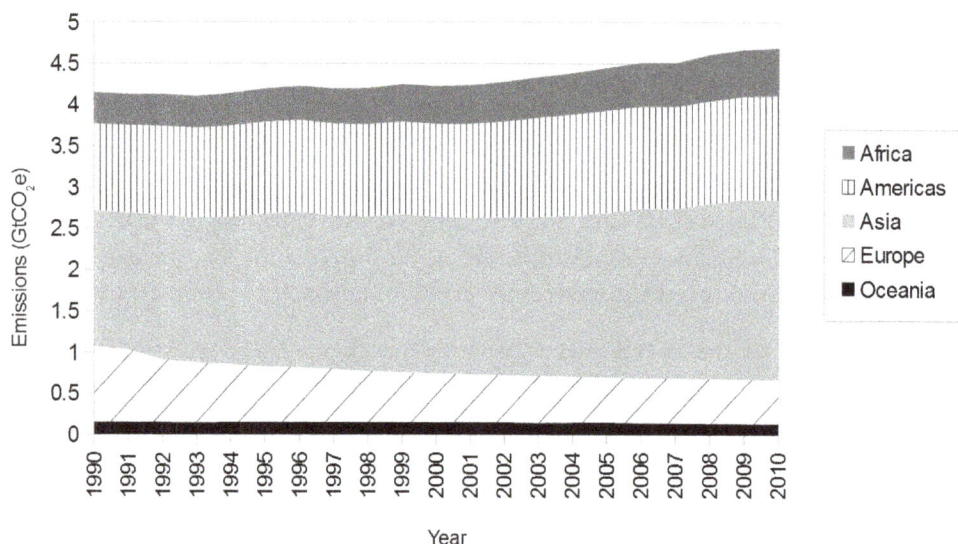

Greenhouse gas emissions from agriculture, by region, 1990-2010.

The agricultural sector is a driving force in the gas emissions and land use effects thought to cause climate change. In addition to being a significant user of land and consumer of fossil fuel, agriculture contributes directly to greenhouse gas emissions through practices such as rice production and the raising of livestock; according to the Intergovernmental Panel on Climate Change, the three main causes of the increase in greenhouse gases observed over the past 250 years have been fossil fuels, land use, and agriculture.

Land Use

Agriculture contributes to greenhouse gas increases through land use in four main ways:

- CO_2 releases linked to deforestation
- Methane releases from rice cultivation
- Methane releases from enteric fermentation in cattle
- Nitrous oxide releases from fertilizer application

Together, these agricultural processes comprise 54% of methane emissions, roughly 80% of nitrous oxide emissions, and virtually all carbon dioxide emissions tied to land use.

The planet's major changes to land cover since 1750 have resulted from deforestation in temperate regions: when forests and woodlands are cleared to make room for fields and pastures, the albedo of the affected area increases, which can result in either warming or cooling effects, depending on local conditions. Deforestation also affects regional carbon reuptake, which can result in increased concentrations of CO_2, the dominant greenhouse gas. Land-clearing methods such as slash and burn compound these effects by burning biomatter, which directly releases greenhouse gases and particulate matter such as soot into the air.

Livestock

Livestock and livestock-related activities such as deforestation and increasingly fuel-intensive farming practices are responsible for over 18% of human-made greenhouse gas emissions, including:

- 9% of global carbon dioxide emissions
- 35–40% of global methane emissions (chiefly due to enteric fermentation and manure)
- 64% of global nitrous oxide emissions (chiefly due to fertilizer use.)

Livestock activities also contribute disproportionately to land-use effects, since crops such as corn and alfalfa are cultivated in order to feed the animals.

In 2010, enteric fermentation accounted for 43% of the total greenhouse gas emissions from all agricultural activity in the world. The meat from ruminants has a higher carbon equivalent footprint than other meats or vegetarian sources of protein based on a global meta-analysis of lifecycle assessment studies. Methane production by animals, principally ruminants, is estimated 15-20% global production of methane.

Worldwide, livestock production occupies 70% of all land used for agriculture, or 30% of the land surface of the Earth.

References

- Wezel, A., Bellon, S., Doré, T., Francis, C., Vallod, D., David, C. (2009). Agroecology as a science, a movement or a practice. A review. Agronomy for Sustainable Development (published online)

- Pretty, Jules. 2008. Agricultural sustainability: concepts, principles and evidence. Philosophical Transactions of the Royal Society, 363, 447-465.

- Dalgaard, Tommy, and Nicholas Hutchings, John Porter. "Agroecology, Scaling and Interdisciplinarity." Agriculture Ecosystems and Environment 100(2003): 39-51.

- *Francis; et al. (2003). "Agroecology: the ecology of food systems". Journal of Sustainable Agriculture. 22 (3): 99–118. doi:10.1300/J064v22n03_10.*

- Buttel, Frederick. "Envisioning the Future Development of Farming in the USA: Agroecology between Extinction and Multifunctionality?" New Directions in Agroecology Research and Education (2003)

- Wojtkowski, Paul A. (2002) Agroecological Perspectives in Agronomy, Forestry and Agroforestry. Science Publishers Inc., Enfield, NH, 356p.

- Boer, I J. M. 2003. Environmental impact assessment of conventional and organic milk production. Livestock Production Science. Vol 80, p 69–77.

- Hovi, M. el al. 2003. Animal health and welfare in organic livestock production in Europe: current state and future challenges. Livestock Production Science. Vol 80, p 41–53.

- Branco, H. and Lal, R. 2008. Principles of Conservation Management. No Tillage-Farming (Ch.8). Springer Verlag. Netherlands. P. 195

- Bolliger, A. et al. 2006. Taking stock of the Brazilian "Zero Till Revolution": A review of landmark research and farmers' practice. Adv. Agron. Vol 91, p 47–110

- Calegari, A. et al. 2008. Impact of Long-Term No-Tillage and Cropping System Management on Soil Organic Carbon in an Oxisol: A Model for Sustainability. Agronomy Journal. Vol 100, Issue 4, p 1013-1019

- West, T. and Post, W. 2002. Soil Organic Carbon Sequestration Rates by Tillage and Crop Rotation: A Global Data Analysis. Soil Sci. Soc. Am. J. 66:1930–1946

- Machado, P.L.O.A. and Silva, C.A. 2001. Soil management under no-tillage systems in the tropics with special reference to Brazil. Nutr. Cycling Agroecosyst. Vol 61, p 119–130

- Koga, N. et al. 2003. Fuel consumption-derived CO_2 emissions under conventional and reduced tillage cropping systems in northern Japan. Agriculture, Ecosystems and Environment. Vol 99, p 213–219.

- Pavuk, D.M. 1994. Influence of weeds within Zea mays crop plantings on populations of adult Diabrotica barberi and Diabrotica virgifera virgifera. Agriculture, Ecosystems & Environment. Vol 50, p 165-175

Permissions

Index

www.ingramcontent.com/pod-product-compliance
Lightning Source LLC
Chambersburg PA
CBHW061318190326
41458CB00011B/3835